Theory of
Partial Differential Equations

This is Volume 93 in
MATHEMATICS IN SCIENCE AND ENGINEERING
A series of monographs and textbooks
Edited by RICHARD BELLMAN, *University of Southern California*

The complete listing of books in this series is available from the Publisher
upon request.

Theory of
Partial Differential Equations

H. MELVIN LIEBERSTEIN

Department of Mathematics
University of Newcastle
Newcastle, New South Wales
Australia

ACADEMIC PRESS   New York and London   1972

211567

ACADEMIC PRESS, INC.
111 Fifth Avenue, New York, New York 10003

United Kingdom Edition published by
ACADEMIC PRESS, INC. (LONDON) LTD.
24/28 Oval Road, London NW1

LIBRARY OF CONGRESS CATALOG CARD NUMBER: 72-84278

AMS (MOS) 1970 Subject Classifications: 35-01, 35-02,
32A05, 65-01, 95-02

PRINTED IN THE UNITED STATES OF AMERICA

Contents

PREFACE   xi

PART I.   AN OUTLINE

Chapter 1   **The Theory of Characteristics, Classification, and the Wave Equation in E^2**

    1. D'Alembert Solution of the Cauchy Problem for the Homogeneous Wave Equation in E^2 3
    2. Nomenclature   8
    3. Theory of Characteristics and Type Classification for Equations in E^2 12
    4. Considerations Special to Nonlinear Cases   17
    5. Compatibility Relations and the Finite-Difference Method of Characteristics   18
    6. Systems Larger Than Two by Two   21
    7. Flow and Transmission Line Equations   22

Chapter 2   **Various Boundary-Value Problems for the Homogeneous Wave Equation in E^2**

    1. The Cauchy or Initial-Value Problem   29
    2. The Characteristic Boundary-Value Problem   29
    3. The Mixed Boundary-Value Problem   32
    4. The Goursat Problem   33

v

5.	The Vibrating String Problem	36
6.	Uniqueness of the Vibrating String Problem	40
7.	The Dirichlet Problem for the Wave Equation?	42

Chapter 3 Various Boundary-Value Problems for the Laplace Equation in E^2

1.	The Dirichlet Problem	45
2.	Relation to Analytic Functions of a Complex Variable	47
3.	Solution of the Dirichlet Problem on a Circle	50
4.	Uniqueness for Regular Solutions of the Dirichlet and Neumann Problem on a Rectangle	51
5.	Approximation Methods for the Dirichlet Problem in E^2	53
6.	The Cauchy Problem for the Laplace Equation	56

Chapter 4 Various Boundary-Value Problems for Simple Equations of Parabolic Type

1.	The Slab Problem	59
2.	An Alternative Proof of Uniqueness	61
3.	Solution by Separation of Variables	62
4.	Instability for Negative Times	62
5.	Cauchy Problem on the Infinite Line	63
6.	Unique Continuation	65
7.	Poiseuille Flow	66
8.	Mean-Square Asymptotic Uniqueness	69
9.	Solution of a Dirichlet Problem for an Equation of Parabolic Type	71

Chapter 5 Expectations for Well-Posed Problems

1.	Sense of Hadamard	73
2.	Expectations	75
3.	Boundary-Value Problems for Equations of Elliptic–Parabolic Type	82
4.	Existence as the Limit of Regular Solutions	84
5.	The Impulse Problem as a Prototype of a Solution in Terms of Distributions	85
6.	The Green Identities	87
7.	The Generalized Green Identity	89
8.	$\mathscr{L}^p$-Weak Solutions	91
9.	Prospectus	93
10.	The Tricomi Problem	94

PART II. SOME CLASSICAL RESULTS FOR NONLINEAR EQUATIONS IN TWO INDEPENDENT VARIABLES

Chapter 6   Existence and Uniqueness Considerations for the Nonhomogeneous Wave Equation in E^2

1.	Notation	101
2.	Existence for the Characteristic Problem	102
3.	Comments on Continuous Dependence and Error Bounds	110
4.	An Example Where the Theorem as Stated Does Not Apply	110
5.	A Theorem Using the Lipschitz Condition on a Bounded Region in E^5	112
6.	Existence Theorem for the Cauchy Problem of the Nonhomogeneous (Nonlinear) Wave Equation in E^2	114

Chapter 7   The Riemann Method

1.	Three Forms of the Generalized Green Identity	118
2.	Riemann's Function	120
3.	An Integral Representation of the Solution of the Characteristic Boundary-Value Problem	124
4.	Determination of the Riemann Function for a Class of Self-Adjoint Cases	126
5.	An Integral Representation of the Solution of the Cauchy Problem	128

Chapter 8   Classical Transmission Line Theory

1.	The Transmission Line Equations	131
2.	The Kelvin r–c Line	133
3.	Pure l–c Line	136
4.	Heaviside's r–c–l–g Distortion-Free Balanced Line	137
5.	Contribution of Du Boise-Reymond and Picard to the Heaviside Position	139
6.	Realization	140
7.	Neurons	141

Chapter 9   The Cauchy–Kovalevski Theorem

1.	Preliminaries; Multiple Series	142
2.	Theorem Statement and Comments	145
3.	Simplification and Restatement	148
4.	Uniqueness	149
5.	The first Majorant Problem	150

CONTENTS

6. An Ordinary Differential Equation Problem   151
7. Remarks and Interpretations   153

PART III. SOME CLASSICAL RESULTS FOR THE LAPLACE AND
WAVE EQUATIONS IN HIGHER-DIMENSIONAL SPACE

Chapter 10   A Sketch of Potential Theory

1. Uniqueness of the Dirichlet Problem Using the
   Divergence Theorem   159
2. The Third Green Identity in E^3 160
3. Uses of the Third Identity and Its Derivation for
   $E^n, n \neq 3$ 165
4. The Green Function   166
5. Representation Theorems Using the Green Function   167
6. Variational Methods   169
7. Description of Torsional Rigidity   170
8. Description of Electrostatic Capacitance, Polarization,
   and Virtual Mass   171
9. The Dirichlet Integral as a Quadratic Functional   172
10. Dirichlet and Thompson Principles for Some Physical
    Entities   174
11. Eigenvalues as Quadratic Functionals   175

Chapter 11   Solution of the Cauchy Problem for the Wave
Equation in Terms of Retarded Potentials

1. Introduction   177
2. Kirchhoff's Formula   178
3. Solution of the Cauchy Problem   183
4. The Solution in Mean-Value Form   185
5. Verification of the Solution of the Homogeneous Wave
   Equation   186
6. Verification of the Solution to the Homogeneous
   Boundary-Value Problem   187
7. The Hadamard Method of Descent   189
8. The Huyghens Principle   193

PART IV. BOUNDARY-VALUE PROBLEMS FOR EQUATIONS OF
ELLIPTIC–PARABOLIC TYPE

Chapter 12   A Priori Inequalities

1. Some Preliminaries   201
2. A Property of Semidefinite Quadratic Forms   203

3. The Generalized Green Identity Using $v = (u^2 + \delta)^{p/2}$ 204
4. A First Maximum Principle   207
5. A Second Maximum Principle   210

Chapter 13   **Uniqueness of Regular Solutions and Error Bounds in Numerical Approximation**

1. A Combined Maximum Principle   215
2. Uniqueness of Regular Solutions   216
3. Error Bounds in Maximum Norm   216
4. Error Bounds in L^p-Norm   218
5. Computable Bounds for the L^2-Norm of an Error Function   219

Chapter 14   **Some Functional Analysis**

1. General Preliminaries   221
2. The Hahn–Banach Theorem, Sublinear Case   225
3. Normed Spaces and Continuous Linear Operators   230
4. Banach Spaces   233
5. The Hahn–Banach Theorem for Normed Spaces   235
6. Factor Spaces   238
7. Statement (Only) of the Closed Graph Theorem   239

Chapter 15   **Existence of $\mathscr{L}^p$-Weak Solutions**

1. A First Form of the Abstract Existence Principle   240
2. Function Spaces $\mathscr{L}^p$ and $\mathscr{L}^{p/(p-1)}$; Riesz Representation   244
3. A Reformulation of the Abstract Existence Principle   245
4. Application of the Reformulated Principle to $\mathscr{L}^p$-Weak Existence   246
5. Uniqueness of $\mathscr{L}^p$-Weak Solutions   248
6. Prospectus   249

NOTES   253

REFERENCES   264

INDEX   267

Preface

This book is written in four modular parts intended as easy steps for the student. The intention here is to lead him from an elementary level to a level of modern analysis research. Thus the first pages of Part I are an explanation of the regular (classical) solutions of the second-order wave equation in two-space–time, while the latter pages of Part IV encompass a more or less complete analysis of the existence of $\mathscr{L}^p$-weak solutions for boundary-value problems for equations of elliptic–parabolic type expounded according to G. Fichera of the University of Rome. In the developing process, an effort is made to ensure that the student samples the extensive variety of mathematically conceivable boundary-value problems, even if their properties are not entirely satisfying once analyzed, and that he learns how to use these tools to elucidate phenomena of nature and technology.

The field is to some extent characterized by the fact that one rarely "solves" boundary-value problems in any acceptable sense of the word. Since computing plays an altogether inseparable role in approximating solutions to boundary-value problems, we present wherever possible a skeleton of the basic theoretical framework for the numerical analysis of several problems along with that of the theory of existence, uniqueness, and integral representations. Where numerical techniques are thought to be suggestive, we present them before presenting existence–uniqueness theories; sometimes when useful and not grossly misleading, we may even present them in lieu of existence–uniqueness theories. Also, we occasionally interrupt other presentations to give some theoretical background of basic computational procedures. However, any serious presentation of the theory of computation procedures is beyond the scope of this book. Nevertheless, we still have tried to present a text in which there is a natural integration of the topics of existence, uniqueness, approximation, and some analysis of computation procedures with applications.

Actually, our purpose has been to write a readable and teachable general text of modern mathematical science—one without substantial pretext to technical originality and yet one that is exciting and thorough enough to provide a basic background. The advantage of the modular approach is that a student may start where he finds his level, stop where his interests stop, and continue at his own rate, even piecemeal if he is so inclined. Courses can easily be organized from the text in the same manner. An instructor will find that he can easily addend or delete material without destroying the continuity of presentation. We believe, in fact, that most instructors want a text that will help them to organize their own course rather one that demands a specific approach. From our experience, we can recommend the following organizations of courses from this text, but we hope other instructors will find their own useful combinations of material and perhaps insert their own favorite topics:

Mode 1: Part I, only—a one-quarter course for students of engineering and physics,

Mode 2: Parts I and II—a one-semester, first-year graduate or senior-level course for students of mathematics, engineering, and physics,

Mode 3: Parts I–III—a two-quarter, first-year graduate or senior-level course for students of mathematics, engineering, and physics,

Mode 4: Parts I–IV—a two-semester, or three-quarter, first-year graduate or senior-level course for students of mathematics, engineering, and physics,

Mode 5: Parts I–IV—a two-quarter (only) course at the third-year graduate level in mathematics (at this level, portions that review functional analysis, for example, can be skipped).

We have found it more pretentious than useful to present here a résumé of Lebesgue integration theory, but have, nevertheless, included a treatment of functional analysis that is fairly complete up to the point where it is required for our presentations. We believe the treatment is as brief and readable as one can find useful in the field. Except for our failure to present Lebesgue integration theory, which is really needed only in the last chapter (and even that can be bravely faced without it), we have kept our prerequisites down to just *some* introductory analysis beyond the level of the usual elementary calculus course and some elementary linear algebra. However, the instructor operating in Mode 1 may choose to dispense with even this requirement. I have taught these materials in Modes 1, 2, and 4. Professor Charles Bryan of the University of Montana, who as a doctoral student under my direction helped to write Chapter 15, has successfully used the materials from this text in Mode 5 (during 1969–1970). His assistance with the writing of this book has been most valuable, and Mode 5 is his idea. We must also acknowledge the influence of Professor Bernard Marcus of San Diego State College in the functional analysis chapter and Professor Robert Stevens of Montana University for a certain example we have used in Chapter 6; at the time when their contributions were made, both were engaged in doctoral studies under my direction.

The Mode 3 presentation does not involve Lebesgue integration beyond the level of some comments, mostly restricted to notes collected at the end of the text.

We have conceived Part I to be our "outline." Here we encourage the student to seek an understanding of the entire field of boundary-value problems by way of a more or less exhaustive study of the simplest linear homogeneous equations of the second order in two independent variables. This material must be completely understood before passing on to the study of the intensively analytic theorems of extensive generality that we find characterizes mature knowledge in this field. Our "outline" material is intended to provide breadth, not depth which, from our point of view, can only come in stages. Simply, successive parts of the book are designed to help in achieving successive levels.

Part II treats of existence and uniqueness by way of Picard iteration of the charteristic and Cauchy (initial value) problems for the wave equation in E^2 with its nonhomogeneous part depending, in a possibly nonlinear way, on the solution and its first partial derivatives. The Riemann method is developed, giving nonsingular integral representations for the linear case. This would seem to represent one of the admirable direct achievements of classical analysis, apparently motivated by Riemann's desire to understand flows of materials under large impact loadings and almost immediately applied by others to achieve an understanding of balanced transmission lines, foreshadowing the advent of clear long distance voice telephony. Transmission lines are studied in a separate chapter in Part II. Part II also treats the Cauchy–Kovalevsky theorem, which concerns analytic solutions of analytic equations corresponding to analytic data on a segment of an analytic initial value curve. The setting of Part II is thoroughly classical in conception throughout. It is quite important and unavoidably difficult in spots, even though it involves no advanced prerequisites.

In Part III we first sketch classical potential theory,* including the usual integral representations for the solution of the Dirichlet problem in terms of the Green function in n dimensions and a somewhat modern approach to variational principles for estimating quadratic functionals. The latter includes studies of such diverse topics as torsional rigidity and bounds for eigenvalues associated with some of the important boundary-value problems. It is then possible to move with ease to a study of the wave equation in higher dimensions, where the intriguing beauty of the Huyghens principle is emphasized and its inner workings exposed by using the Hadamard method of descent. Classical analysis eventually became heavily burdened with clever but extensive and delicate manipulations—presumably this overburden on analysis caused functional analysis to be conceived—and the latter portions of Part III unavoidably reflect this heavy manipulative style, but again it involves no advanced prerequisites.

Part IV presents a résumé of functional analysis, develops several a priori estimates for equations of elliptic–parabolic type (second order, n dimensions) from the

---

*This topic was once a semester or even year graduate course in mathematics.

divergence theorem, uses these to prove uniqueness of certain boundary-value problems, generates bounds therefrom for errors of approximation, and finally develops an **Abstract Existence Principle**, which is used to prove the existence of $\mathscr{L}^p$-weak solutions. This last assumes, but does not require, a rudimentary knowledge of Lebesgue integration theory. *Part IV is intended to provide in very specific terms a picture of the general techniques now being used in modern studies of partial differential equations.* At the end of Part IV, a discourse is undertaken concerning various modern senses of existence. Their physical relevance is reviewed, perhaps too briefly, but to the best of our ability. This is found to suggest a *possibility* that it is *perhaps* primarily the sense of uniqueness that we should think now to weaken—perhaps we should weaken it to time-asymptotic uniqueness with a quickly acquired (unique) steady state—retaining our classical (regular) sense of existence and at the same time insisting that all applied problems be treated as time dependent and not as stationary. If there is any technical, as opposed to expository, originality to be claimed for this text, it is the development of this thesis. We have tried, however, not to impose a private view onto a public body, simply asking that an awareness of such issues and an open mind concerning them be maintained. These, after all, are the issues raised in the last 20 years of progress in partial differential equations, and the effect of these 20 years has been so profound that the thinking in the field will never be the same again.

Perhaps the field of partial differential equations has suffered from too intense specialization among its adherents in the last several generations, but the danger now is too much generality taken on too fast by students without sufficient grounding in "real problems." We have tried here to introduce increased generality at a modest rate of increasing abstraction in stages that would seem to develop its justification in terms of problems that appear to be "real" at each stage.

Notes of general scientific and historical interest are collected at the end of the book (keyed to sections of various chapters) in order not to interrupt the flow of mathematical developments.

An Outline

The Theory of Characteristics, Classification, and the Wave Equation in E^2

1 D'ALEMBERT SOLUTION OF THE CAUCHY PROBLEM FOR THE HOMOGENEOUS WAVE EQUATION IN E^2

Let us consider under what conditions it is possible to determine a unique solution of the equation

$$u_{xx} - u_{yy} = 0 \tag{1.1.1}$$

satisfying the conditions

$$u(x,0) = f(x) \quad \text{and} \quad u_y(x,0) = g(x) \tag{1.1.2}$$

where $f: (a, b) \to R^1$ and $g: (a, b) \to R^1$. We understand that as part of this task we are to decide precisely what we wish to mean by saying a function u is a solution of (1.1.1), (1.1.2) and what properties given functions f and g must have so that such a solution exists and is unique. Toward this end we rewrite Eq. (1.1.1) in coordinates rotated through 45°,

$$\xi = \tfrac{1}{2}(x + y), \qquad \eta = \tfrac{1}{2}(x - y), \tag{1.1.3}$$

by the use of the chain rule on the function u. Let us remark here once and for all that there are really two functions involved, one a function of x and y, another a function of ξ and η formed as a composite of the first with (1.1.3), both of which have the same functional values at those points (x, y) and (ξ, η)

3

which are identified by (1.1.3). Because of this sameness of functional values arising by the formation of composite functions, we will use the same symbol u for both functions. Thus we may write $u(x, y)$ or $u(\xi, \eta)$ for functional values if the distinction of which function is used is not otherwise clear by the context, but the one symbol u will be used for both functions. Far from leading to confusion, as long as we agree to what is being done, this will help keep our bookkeeping straight as to which functional values are to be identified. This will be especially useful if we encounter long strings of changes of variables as one very often does in extensive application areas. As far as we know, all textbooks in partial differential equations are written using this convention, but in these times when the distinction between functions and their functional values is being greatly emphasized even in elementary training it would seem to need statement. In any case, it will be used throughout this text and not mentioned again unless clarification seems specifically demanded by the nature of the arguments presented.

From the chain rule we have

$$u_x = u_\xi \xi_x + u_\eta \eta_x = \tfrac{1}{2}(u_\xi + u_\eta)$$

$$u_y = u_\xi \xi_y + u_\eta \eta_y = \tfrac{1}{2}(u_\xi - u_\eta)$$

$$u_{xx} = \tfrac{1}{2}[(u_\xi + u_\eta)_\xi \xi_x + (u_\xi + u_\eta)_\eta \eta_x]$$

$$= \tfrac{1}{4}(u_{\xi\xi} + u_{\eta\eta}) + \tfrac{1}{2}u_{\xi\eta}$$

$$u_{yy} = \tfrac{1}{2}[(u_\xi - u_\eta)_\xi \xi_y + (u_\xi - u_\eta)_\eta \eta_y]$$

$$= \tfrac{1}{4}(u_{\xi\xi} + u_{\eta\eta}) - \tfrac{1}{2}u_{\xi\eta}$$

so that (1.1.1) becomes

$$u_{\xi\eta} = 0. \tag{1.1.4}$$

Here it has been assumed that $u_{\xi\eta}$ and $u_{\eta\xi}$ are continuous and are therefore equal; i.e., we are now restricted to seek a solution with this property.

We now seek the class of all solutions of (1.1.4). Equation (1.1.4) implies that u_ξ is a function of ξ alone. If this function is integrable, then we may write

$$u(\xi, \eta) = F(\xi) + G(\eta) \tag{1.1.5}$$

where G is an arbitrary function of η introduced by this last "integration" and F, being the primitive of u_ξ (a function of ξ alone), is also arbitrary.

We have thus shown that all solutions of (1.1.4) such that u_ξ is integrable are of the form (1.1.5). Now we must ask if all forms (1.1.5) are solutions of

(1.1.4). The question resolves to, what do we mean by a solution? Here we simply ask that all terms specifying quantities in (1.1.4) exist in some region R where this question is to be resolved and that (1.1.4) be satisfied in that region. But since we ask for an equality to be satisfied, we will also ask that all terms in the equation be continuous—here that $u_{\xi\eta}$ be continuous in the region of consideration. It is evident that the function u as given in (1.1.5) is a solution in this very concrete sense if $F, G \in C^1$ on sufficiently large open intervals of ξ and η.

Reverting back to the original coordinates, we have from (1.1.5)

$$u(x, y) = F(x+y) + G(x-y), \tag{1.1.6}$$

where the definitions of F and G have been altered to absorb the factor $\frac{1}{2}$ in the arguments. Here u is a solution of (1.1.1) in the sense that all terms exist and are continuous in a region of consideration if $F, G \in C^2$, and, moreover, if $F, G \in C^2(a, b)$, then one can see that $u \in C^2(T)$ where T is an isosceles triangle built on the base (a, b). Clarification of the latter will be undertaken in a moment; for now, we should note that the properties required of F and G in order that u in (1.1.5) be a solution in (ξ, η) coordinates are somewhat weaker than the properties required of them in order that (1.1.6) be a solution in (x, y) coordinates. This is a peculiar property of solutions of partial differential equations when considered in this very direct concrete sense, and it is one of the reasons (not the most cogent, however) why many modern workers prefer a more abstract sense of the existence of solutions. Such workers will be seen to lose much, however, in the way of useful physical interpretations of their results when they weaken the sense of existence. A balanced consideration of whether one should use the concrete sense (regular solutions, as we call them) or an abstract sense (e.g., $\mathscr{L}^p$-weak solutions) of the existence of solutions is a theme that will be threaded through this text but it has little relevance yet, and at first we are compelled to consider only the concrete sense of solutions. To some extent, where possible, it will be found that we prefer to weaken the sense of uniqueness rather than existence. Again that is far ahead of the story.

To find F and G in (1.1.6) so that (1.1.2) is satisfied we put

$$f(x) = F(x) + G(x) \tag{1.1.7}$$

and

$$g(x) = F'(x) - G'(x) \tag{1.1.8}$$

where $F(x+y)$ and $G(x-y)$ have been differentiated as composite functions of x and y and then y has been put equal to zero. Letting c be any real number, and assuming g is integrable, (1.1.8) can be written

$$\int_c^x g(s)\,ds = F(x) - G(x). \tag{1.1.9}$$

Then from (1.1.7) and (1.1.9),

$$F(x) = \frac{1}{2}\left[f(x) + \int_c^x g(s)\,ds \right] \tag{1.1.10}$$

and

$$G(x) = \frac{1}{2}\left[f(x) - \int_c^x g(s)\,ds \right]. \tag{1.1.11}$$

These are functions of one variable, but this one variable appears as functional values of two different functions of two variables in (1.1.6), both as $x+y$ and $x-y$. With this in mind we see that (1.1.6) becomes

$$u(x,y) = \tfrac{1}{2}[f(x+y) + f(x-y)] + \tfrac{1}{2}\int_{x-y}^{x+y} g(s)\,ds \tag{1.1.12}$$

where it is also seen that the arbitrary reference value c no longer appears.

This, i.e. (1.1.12), is what we refer to as the D'Alembert solution. D'Alembert is one of our classical fountainheads, so this solution is hardly recent. It provides us with a starting or reference point from which to depart for an understanding of many things. But is it a solution in our concrete sense, and if so, in what region? One sees immediately that (1.1.2) is satisfied and that (1.1.12) is of the form (1.1.6); a quick glance shows that u can be twice continuously differentiated wherever f can be and where g can be once continuously differentiated.

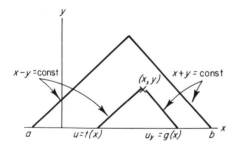

FIG. 1   The solution is uniquely determined in the triangle bounded by $y = 0$, $x - y = $ const, and $x + y = $ const.

Let us select a point (x, y) (see Fig. 1) and ask about the value of u at this point. Draw a line through this point so that $x + y$ is constant and another so that $x - y$ is constant, and note where these lines cross the x axis. There we pick up the values $f(x + y)$ and $f(x - y)$ to use in (1.1.12). Also, the integral term in (1.1.12) is just the integral of g between these points of intersection. From Fig. 1, then, with the comments in the paragraph above, it becomes clear that D'Alembert solution (1.1.12) is indeed a solution in our concrete (regular) sense in the 45° isosceles triangle T with base (a, b) if $f \in C^2(a, b)$ and $g \in C^1(a, b)$.

But is the D'Alembert solution a unique regular solution? Before undertaking this question let us record just what it is that we have now decided to call a regular solution: Let T be an open 45° isosceles triangle with base (a, b) on the x axis. If $u: T \to R^1$ and

(i) $u \in C^2(T)$;

(ii) $u \in C(T \cup (a, b))$;

(iii) (1.1.1) is satisfied for every $(x, y) \in T$; and

(iv) (1.1.2) is satisfied for every $x \in (a, b)$;

then u is said to be a regular solution of (1.1.1), (1.1.2) in T. Condition (ii) provides a connection between the specification of the data as required by (iv) on $y = 0$ and the specification of the differential equation as required by (iii) in the (open) region† T. Some such condition will always be needed in the specification of a boundary-value problem as one can see, but it will sometimes be weakened when the boundary-value problem is restated as an integral equation, and we will often strengthen it to $u \in C^1$ in order to use the divergence theorem conveniently in this outline.

Uniqueness is handled here, as it will always be handled for linear (see the next section) problems, by a simple contradiction argument. Suppose there are two regular solutions u_1 and u_2 of (1.1.1), (1.1.2). Then

$$u = u_1 - u_2$$

satisfies the homogeneous equation

$$u_{xx} - u_{yy} = 0 \tag{1.1.13}$$

and the homogeneous data

$$u(x, 0) = u_y(x, 0) = 0. \tag{1.1.14}$$

$\dagger$ We will always mean by a region an open, connected set.

We must show that (1.1.13) and (1.1.14) together imply that $u = 0$ for every $(x, y) \in T$ so that u_1 and u_2 are equal on T. The problem (1.1.13), (1.1.14), or an appropriate adaptation of it, is always the uniqueness problem for linear equations, and we will not always feel compelled to mention this oft-repeated argument when it is being repeatedly used.

Here for the uniqueness problem we have the opportunity, exceedingly rare in partial differential equations, to use the form (1.1.6) giving all regular solutions. Obviously, (1.1.13), (1.1.14) require that $F = G = 0$ in (1.1.6) and uniqueness is established.

The triangle T is called the "region of determination" of the interval (a, b). The "domain of dependence" D of a point (x, y) is the base of an isosceles triangle on the x axis with (x, y) as its apex. The "region of influence" R of the interval (a, b) is the infinite region shown in Fig. 2 between $x - y = b$ and $x + y = a$. From (1.1.12) and the arguments given above, the student will readily agree that these entities are well named.

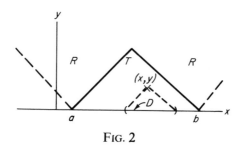

FIG. 2

2 NOMENCLATURE

The term "linear" was used in connection with our discussion of uniqueness before its meaning was stated here. To avoid such occurrences, we now interrupt our substantive presentation to display some of the basic nomenclature in the field.

Let $f: R^8 \to R^1$ (or possibly $f: C^8 \to C^1$). Then a second-order partial differential equation for a function $u: R^2 \to R^1$ (or possibly $u: C^2 \to C^1$) is an equation

$$f(x, y, u, u_x, u_y, u_{xx}, u_{xy}, u_{yy}) = 0. \tag{1.2.1}$$

A definition for a higher-order (referring to the highest number of derivatives of u appearing) equation and one involving more independent variables, $u: R^n \to R^1$ for $n > 2$, can easily be rendered by the student. Invariably some

8

conditions on *f* will be included in the statement of any particular problem or theorem; it may be required that it be possible to solve (1.2.1) for u_{yy} or that *f* be linear in u_{xx}, u_{xy}, u_{yy} or even that *f* be linear in all but the first two entries, *x* and *y*.

If *f* is linear in the highest-order derivatives appearing, then the equation (1.2.1) is described as "quasi linear." In this case (1.2.1) can be written

$$a(x, y, u, u_x, u_y) u_{xx} + 2b(x, y, u, u_x, u_y) u_{xy} + c(x, y, u, u_x, u_y) u_{yy}$$
$$= d(x, y, u, u_x, u_y), \tag{1.2.2}$$

where $a, b, c, d: R^5 \to R^1$. The left member, the sum of highest-order terms appearing, is called the principle part. This part will be found to play an important role, telling us what curves are characteristic, as introduced in the next section, and, therefore, what kinds of boundary-value problems are proper and in what regions solutions are uniquely determined.

If *f* is linear in *u* and all its derivatives, then the differential equation is said to be linear. In this case (1.2.1) can be written

$$a(x, y) u_{xx} + 2b(x, y) u_{xy} + c(x, y) u_{yy}$$
$$= \alpha(x, y) + \beta(x, y) u + \gamma(x, y) u_x + \delta(x, y) u_y \tag{1.2.3}$$

where $a, b, c, \alpha, \beta, \gamma, \delta: R^2 \to R^1$. If $a, b, c, \beta, \gamma, \delta \in R^1$ (i.e., $w_1, w_2, w_3, w_4, w_5, w_6$ $\in R^1$ and $a: R^1 \to \{w_1\}$, $b: R^1 \to \{w_2\}$, $c: R^1 \to \{w_3\}$, $\beta: R^1 \to \{w_4\}$, $\gamma: R^1 \to \{w_5\}$, $\delta: R^1 \to \{w_6\}$), then (1.2.3) is said to be "of constant coefficients." If $\alpha(x, y) = 0$ for every (x, y) (in some region *R* of our consideration) then (1.2.3) is said to be homogeneous; here $u = 0$ is a solution. The wave equation (1.1.1) is linear, homogeneous, and of constant coefficients. Its principle part is called "the wave operator" and constitutes all nonzero terms of the equation. We will find in Section 3 that the equation is "of hyperbolic type" and the two families of "characteristics" are given by $x \pm y = $ constant, the curves that were found to bound the region of determination in Section 1. Moreover, we will find that this was no accident but is to be expected as a general property of hyperbolic type, and this serves to distinguish hyperbolic type. Of course, in general, equations of constant coefficients are by far the easiest to understand, and in large measure our considerations of boundary-value problems in this outline will be for equations of constant coefficients, mostly homogeneous ones. The linear case is much the easier to work with in theoretical questions because the principle of superposition applies for the homogeneous linear case: We leave it to the student to show that if u_1 and u_2 are solutions of the homogeneous equation (1.2.3) [i.e., with $\alpha(x, y) = 0$ for every (x, y) in the region of consideration], then for $m, n \in R^1$, $mu_1 + nu_2$ is a solution of the same equation. This is the all-important principle of superposition; really it, rather than that f is linear, should be thought of as characterizing linear equations.

Let $F_i: R^{3n+2} \to R^1$, $i = 1, ..., n$. Then the $n \times n$ system of partial differential equations

$$F_1(x, y; u_1, ..., u_n; u_{1,x}, ..., u_{n,x}; u_{1,y}, ..., u_{n,y}) = 0$$
$$\vdots \qquad\qquad (1.2.4)$$
$$F_n(x, y; u_1, ..., u_n; u_{1,x}, ..., u_{n,x}; u_{1,y}, ..., u_{n,y}) = 0$$

is said to be a first-order system of partial differential equations in n real-valued functions $u_i: R^2 \to R^1$ of two real variables x, y. *For purposes of simplicity, we will often speak of the $n = 2$ case,* though this will often have to be followed by a discussion of complications arising in the general case. When no such discussions follow, unless we are discussing a very specific equation, the obvious generalizations apply and the $n = 2$ case is stated as a prototype.

Following this approach, we now look at the quasi-linear case where, of course, the functions F_i, $i = 1, ..., n$, are linear in $u_{i,x}$, and $u_{i,y}$, $i = 1, ..., n$; in the $n = 2$ case these equations are

$$a_{11} u_x + a_{12} u_y + b_{11} v_x + b_{12} v_y = h_1$$
$$a_{21} u_x + a_{22} u_y + b_{21} v_x + b_{22} v_y = h_2 \qquad (1.2.5)$$

where $a_{ij}, b_{ij}, h_j: R^{m+2} \to R^1$. In the $n \times n$ case we can write

$$AU_x + BU_y = H \qquad (1.2.6)$$

where $A = (a_{ij}(x, y, u_1, ..., u_n))$ and $B = (b_{ij}(x, y, u_1, ..., u_n))$ are $n \times n$ square matrices of real-valued functions and $H = (h_j(x, y, u_1, ..., u_n))$ is a column matrix of real-valued functions. Of course, once again, if functions h_j, $j = 1, ..., n$, are linear in $u_1, ..., u_n$ and A and B are functions of (x, y) alone, then (1.2.6) [or (1.2.5) with $n = 2$] is said to be linear; the functions F_j, $j = 1, ..., n$, in (1.2.4) will be linear in functions $u_j, u_{jx}, u_{jy}, j = 1, ..., n$, and the principle of superposition will apply for the homogeneous case. Of course, "homogeneous" is now defined in an obvious way. If A and B are matrices of constants and H is a column matrix of functions

$$\alpha(x, y) + \sum_{i=1}^{m} \beta_i u_i \qquad \text{where} \quad \beta_i \in R^1,$$

then (1.2.6) is said to be "of constant coefficients."

It will be important to notice in Section 3 and following that systems of first-order equations are more general than one higher-order equation. This does not mean that we will always prefer to treat an equation in the form (1.2.5) [or (1.2.6)] rather than in the form (1.2.2), but rather that one *can* always pass from an *n*th-order equation to an $n \times n$ first-order system (1.2.6) if only regular solutions are sought, and *the reverse is not always possible.* We leave it to the student to find a counterexample and verify the latter contention. For the former, we will demonstrate the principle on Eq. (1.2.2). Let

$$\varphi = u_x, \qquad \psi = u_y \tag{1.2.7}$$

where it is assumed that $u \in C^2(R)$ in some region R of our consideration. Then from (1.2.2) we have in R that

$$a\varphi_x + b\varphi_y + b\psi_x + c\psi_y = d$$

and $\hspace{8cm}$ (1.2.8)

$$\varphi_y - \psi_x = 0$$

which is a first-order system equivalent to (1.2.2) in R. It is clearly in the form of (1.2.5) with

$$a_{11} = a, \qquad a_{12} = b_{11} = b, \qquad b_{12} = c, \qquad h_1 = d$$
$$a_{21} = 0, \qquad a_{22} = 1, \qquad b_{21} = -1, \qquad b_{22} = 0, \qquad h_2 = 0.$$

Thus in Section 3 we will classify equations of the form (1.2.5) and (1.2.6), confident that these considerations apply equally well to an equation of the form (1.2.2) or to higher-order ones. In Sections 3 and 7 we will take up specific examples of equations in the form (1.2.2) and consider classifications.

In the above we have mentioned only equations with two independent variables, but the expressions used could easily be generalized to n independent variables by replacing x and y by x_i, $i = 1, 2$, but forming all indicated sums up to $i = n$. Thus, for example, (1.2.3) could be replaced by

$$\sum_{j=1}^{n} \sum_{i=1}^{n} a^{ij} u_{x_i x_j} + \sum_{i=1}^{n} b^i u_{x_i} + cu = \alpha \tag{1.2.9}$$

where $a^{ij}: R \to R^1$, $b^i: R \to R^1$, $c: R \to R^1$, $\alpha: R \to R^1$, and R is some region of consideration. In fact, we will study this general form of a second-order linear equation in n variables (in n-dimensional Euclidean space) but written in the form

$$a^{ij} u_{x_i x_j} + b^i u_{x_i} + cu = \alpha \tag{1.2.9a}$$

where indices repeated in any one term require summation from 1 to n.

11

Naturally, equations in two independent variables are more difficult than those in one; there is less known (except for cases degenerating to one independent variable) and more to know. Much of this outline will be devoted to showing the variety of boundary-value problems that two dimensions allows, an almost unprecedented feature that arises in passing from one to two dimensions. In passing to three independent variables, we will also encounter increased difficulties; but, for theoretical purposes, it will be found that these difficulties do not increase as we pass to $n > 3$-dimensional space. For $n \geqslant 3$, as we increase the dimension by one, we may well encounter greatly increased difficulty in numerical approximation, in the sense that the magnitude of the calculations required may well become so large that one must devise very clever schemes to handle them, but theoretical difficulties are rarely encountered in passing to higher dimension for $n \geqslant 3$.

3 THEORY OF CHARACTERISTICS AND TYPE CLASSIFICATION FOR EQUATIONS IN E^2

We consider the quasi-linear system (1.2.5) as a prototype for the $n \times n$ system (1.2.6) although we agree from the outset that additional comments over those related to (1.2.5) will be necessary in order to accommodate all the many possibilities in (1.2.6). We consider a region D in the (x, y) plane and a curve segment

$$c = \{(x,y)|(y = k(x)) \wedge k : I \to R^1 \wedge k \in C^1(I)\} \subset D,$$

where I is a closed interval on the x axis and the condition $k \in C^1(I)$ prevents there being any vertical tangents to c. Let $f : D \to R^1$ and let $f \in C^1(D)$. Then $f R c$ (f restricted to c) can be regarded as a function of one variable. Following our earlier announced plan to use the same notation for a function and a composite from it, we write

$$df/dx = f_x + k'f_y \qquad (1.3.1)$$

where, of course, on c

$$dy/dx = k'(x). \qquad (1.3.2)$$

We ask the question: *Do there exist segments c such that if u and v are defined on c, the system* (1.2.5) *does not uniquely define* u_y *and* v_y? The question will be answered very specifically by substituting for u_x and v_x in (1.2.5) in terms of u_y and v_y according to (1.3.1), with u and v replacing f. This will give conditions on coefficients a_{ij} and b_{ij} so that such curves do exist, and, moreover, it will provide differential equations for them. Such curves are called "characteristics," and they are very important in understanding the properties of partial differential equations.

Why are they so important? To answer this let us first note that the specification of u and v on c together with Eq. (1.2.5) is the equivalent of the specification of an initial-value problem for one second-order equation. For example, (1.1.1) may, as we have seen, be written as a system by letting u_x and u_y be new dependent variables. Then the specification of $u(x, 0) = f(x)$ can be replaced by $u_x(x, 0) = f'(x)$, and we already have $u_y(x, 0) = g(x)$. The curve segment c becomes simply the interval (a, b) of the x axis. If a more complicated curve c [represented by a differentiable function $y = k(x)$] were used, the chain rule would have to be used to convert the initial-value problem for one second-order equation to that of a 2×2 system, but the same principles would hold.

The importance of our question is that curves which fit the specification are ones for which solutions of the initial-value (or Cauchy) problem are not unique. If differentiable solutions of initial-value (or Cauchy) problems with data u and v on segment c were uniquely determined in a region D containing c, then on differencing values at points on the curve and vertically above or below them and taking limits of difference quotients we could find unique values of u_y and v_y. In Fig. 3, for example, we would simply have with $h = y_2 - y_1$

$$u_y = \lim_{h \to 0} \frac{u_2 - u_1}{h}.$$

We speak of u_y and v_y rather than du/ds and dv/ds, where s is a parameter in any direction off the curve, simply for our own convenience. The condition, as we have noted, that $k \in C^1(I)$ guarantees that y is in a direction not tangent to the curve.

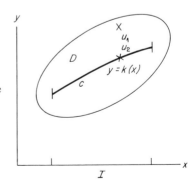

FIG. 3 u_y on c is the limit of a difference quotient from vertical points.

From the above, then, characteristics are curves on which the specification of Cauchy data does not give a unique determination of a solution—the specification of other data, such as the specification of u or v on another intersecting curve might help, but Cauchy data alone are not enough to specify a unique solution when given on a characteristic. This fact bears within it the important feature that it may not always be possible "to continue" solutions uniquely across characteristics. If by some circumstances, one is able to find the solution of a problem up to a characteristic segment but not including it, and if by assuming continuity of the u and v functions only (not their derivatives), one extends the domain of these functions to include this segment, the question arises, can these functions be extended uniquely to a domain beyond the segment so that (1.2.5) is satisfied and u and v are continuous in the whole new domain of their definition? According to the question, and the definition it implies for characteristics, they cannot. Thus characteristic segments, if they exist, will tend to serve as boundaries for regions of determination just as indeed we found that the curves $x \pm y =$ constant did for the problem (1.1.1), (1.1.2). They serve, so to speak, as natural barriers for the unique continuous extension of solutions of a system like (1.2.5).

Let us now pursue the manipulations we have said are attendant upon the question. From (1.3.1) with u and v replacing f,

$$u_x = du/dx - k'u_y \qquad \text{and} \qquad v_x = dv/dx - k'v_y \qquad (1.3.3)$$

which in (1.2.5) gives

$$(a_{12} - k'a_{11})u_y + (b_{12} - k'b_{11})v_y = h_1 - a_{11}\,du/dx - b_{11}\,dv/dx$$
$$(a_{22} - k'a_{21})u_y + (b_{22} - k'b_{21})v_y = h_2 - a_{21}\,du/dx - b_{21}\,dv/dx. \qquad (1.3.4)$$

It is to be noted here that we regard du/dx and dv/dx to be known on c because u and v are known on c and are obviously sought such as to be differentiable (regular solutions) in D. The derivatives again are limits of difference quotients. The question then of whether or not a curve segment c defined by $y = k(x)$, $k \in C^1(I)$, is characteristic is answered by noting whether or not the coefficient matrix in (1.3.4) is singular. That is, k' is a characteristic direction if

$$\begin{vmatrix} a_{12} - k'a_{11} & b_{12} - k'b_{11} \\ a_{22} - k'a_{21} & b_{22} - k'b_{21} \end{vmatrix} = 0 \qquad (1.3.5)$$

because (1.3.4) is simply a linear system in two unknowns, u_y and v_y. We will call (1.3.5) "the characteristic equation." It can have at most two roots k_1', k_2'—real and distinct, real and equal, or complex conjugates (if a_{ij}, b_{ij}

have real values). This provides our major classification of partial differential equations:

hyperbolic type: $\qquad k_1' \in R^1, \quad k_2' \in R^1, \quad k_1' \neq k_2'$

parabolic type: $\qquad k_1' = k_2' \in R^1$ $\qquad\qquad$ (1.3.6)

elliptic type: $\qquad k_1' \in C^1, \quad k_2' \in C^1, \quad k_1' \neq k_2', \quad \bar{k}_1' = k_2'.$

Our discussions above should now give significance to these classifications. Consider for a moment the linear case. Here the a_{ij}'s and b_{ij}'s are functions of x and y alone and so, then, are the roots of (1.3.5). In the hyperbolic case we have the two real equations

$\qquad$ (a) $\quad dy/dx = k_1'(x,y)$ $\qquad$ (b) $\quad dy/dx = k_2'(x,y)$ $\qquad$ (1.3.7)

for the characteristics. Let us examine if characteristics pass through a point $(x_0, y_0) \in D$. Expressed another way, we have that

$$y(x_0) = y_0 \qquad\qquad (1.3.8)$$

as a condition to add both to (1.3.7a) and (1.3.7b). But these then give initial-value problems for first-order ordinary differential equations. If, then, the well-known Lipschitz condition is satisfied on both roots k_1', k_2' of the equation (1.3.5), there is a neighborhood of (x_0, y_0) in which two unique characteristic curves are determined passing through (x_0, y_0). A region D of hyperbolicity is then characteristic rich—anywhere we put a pencil, there pass two characteristics (see Fig. 4)—if the roots of the characteristic equation satisfy a Lipschitz condition (in appropriate regions). We thus have two real families of characteristics which form possible barriers for unique continuous extension and can act as bounding segments for regions of determination.

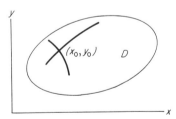

FIG. 4   For an equation hyperbolic in D, two characteristics pass through each point in D.

It should be evident now that for parabolic type there is only one such family, only one boundary—at least not intersecting boundaries—of a region of determination that can be expected to arise naturally without prior specification. For elliptic type, then, there are no such real barriers for unique continuation.

15

We will find that the Laplace equation is of elliptic type and that inside a region where this equation is satisfied, even analytic continuations—ones having convergent power series—are always possible over any curve c. Since no segments in D are characteristic for a system (1.2.5) of elliptic type in D, we can always compute u_y and v_y uniquely on a curve c where u and v are known. But then, knowing these, we can compute their y derivatives, and repeating, we can compute all orders of y derivatives. Since u and v are known (and differentiable) du/dx and dv/dx are known so that by way of (1.3.1), u_x and v_x can be computed as well as all other x and mixed derivatives. Thus power series for solutions of elliptic-type systems (1.2.5) can always be uniquely determined from smooth data u and v on a given (infinitely smooth) curve c. The question of convergence (and radius of convergence) then remains. This question and other similar ones will eventually be answered by the Cauchy–Kovalevski theorem. We have simply stated the answer in two dimensions because the real and imaginary parts of an analytic function of a complex variable satisfies the Laplace equation in R^2, and the fact is a familiar one in elementary complex variables theory.

The Laplace equation

$$u_{xx} + u_{yy} = 0$$

can be written, according to our prescription in (1.2.7) and (1.2.8), in such a way as to get the familiar Cauchy–Riemann equation. The student should now write the Laplace equation, the wave equation, and the heat equation $(u_{xx} - u_y = 0)$ as first-order systems and undertake the following exercises, not formally by using the general formula (1.3.5), but by repeating in each case the logical steps which lead to (1.3.5).

EXERCISE 1 Show that for the wave equation, the real curves $x \pm y = $ constant are the two families of characteristics.

EXERCISE 2 Show that for the Laplace equation, the equations $ix \pm y = $ constant are characteristics, but these are not real curves.

EXERCISE 3 Show that for the heat equation $y = $ constant is the single family of characteristics.

EXERCISE 4 Show that the Tricomi equation, $yu_{xx} + u_{yy} = 0$, is of elliptic type in the upper half-plane, parabolic type on the x axis, and hyperbolic type in the lower half-plane. Find the "semicubical parabola" characteristics in the lower half-plane.

16

EXERCISE 5 From (1.2.8) show that the characteristic directions for the general quasi-linear equation (1.2.2)

$$au_{xx} + 2bu_{xy} + cu_{yy} = d$$

are given by

$$\frac{dy}{dx} = \frac{b \pm (b^2 - ac)^{1/2}}{a}. \tag{1.3.9}$$

From the last exercise it is seen that (1.2.2) is classified according to:

$$\begin{array}{lll} \text{hyperbolic type:} & b^2 - ac > 0 & \\ \text{parabolic type:} & b^2 - ac = 0 & (1.3.10) \\ \text{elliptic type:} & b^2 - ac < 0 & \end{array}$$

This will justify the names in this classification of type if (1.2.2) is regarded as a quadratic form in the partial differential operators $(\ )_x$ and $(\ )_y$. The statement (1.3.10) was once given in all texts as the classification without further justification being offered or seeming to be required.

EXERCISE 6 Using (1.3.9), find the characteristics for the Fichera equation,

$$y^2 u_{xx} - 2xyu_{xy} + x^2 u_{yy} + yu_x + xu_y - u = 0.$$

4 CONSIDERATIONS SPECIAL TO NONLINEAR CASES

If in (1.2.5), the coefficients a_{ij}, b_{ij} $(i, j = 1, ..., m)$ are functions of x, y, u, and v, not simply of x and y as in the linear case, then the differential equations (1.3.7a, b) arising from computing roots of (1.3.5) are

$$\text{(a)} \quad dy/dx = k_1'(x, y, u, v) \qquad \text{(b)} \quad dy/dx = k_2'(x, y, u, v). \tag{1.4.1}$$

The arguments of Section 3 for existence of two characteristics through any point (x_0, y_0) could be carried out if solutions $u = u(x, y)$ and $v = v(x, y)$ were known but obviously not otherwise because there are too many variables involved in the system (1.4.1a, b) if this is all the information to be carried. Thus, if the principle part of an equation or system of equations is nonlinear, a knowledge of characteristics (even their existence) depends on a knowledge of solutions (and their existence). Relative to solutions $u = u(x, y)$ and $v = v(x, y)$, there will exist unique solutions of (1.4.1a, b) passing through a point (x_0, y_0) if, according to the specifications of $u(x, y)$ and $v(x, y)$,

$$k_1'(x, y, u(x, y), v(x, y)) \qquad \text{and} \qquad k_2'(x, y, u(x, y), v(x, y))$$

satisfy an appropriate Lipschitz condition. Conversely, if differentiable characteristics $y = k_1(x)$ and $y = k_2(x)$ are known, then on forming dy/dx for each of them and inserting these into (1.4.1a, b), we obtain a system of two nonlinear equations for u and v in terms of x, y. Then for any (x, y) such that the Jacobian, $J(k_1', k_2'/u, v)$ is not singular, we have that u and v exist. Thus, in general, for equations in which the principle part is nonlinear, the characteristics and the solutions are inextricably tied together, one depending on the other. The characteristics cannot be known "in advance" over a region until the solutions are specified.

What does it take uniquely to specify the solution? In a sense this will be the question to which this entire text will be addressed; it involves the whole question of what type of boundary-value problems are well posed. The picture is not so bleak, however, as it may first appear for useful consideration of the theory of characteristics on equations with nonlinear principle part. Suppose, for example, we wish to consider a Cauchy problem, u and v given on a segment of a curve $y = \gamma(x)$, where $\gamma: (a, b) \to R^1$ and $\gamma \in C^1(a, b)$. According to considerations above concerning the definition of characteristics, we will have to make sure that the initial segment γ is noncharacteristic—i.e., nowhere tangent to a characteristic—or, otherwise, we surely know in advance that the solution is not uniquely determined even in a small neighborhood of points of γ. But on γ we know the solutions u and v—they are specified there—so in order to assure ourselves that γ is noncharacteristic we simply check that

$$\gamma' \neq k'_{1,2}(x, \gamma(x), u(x, \gamma(x)), v(x, \gamma(x))). \qquad (1.4.2)$$

In summary, the lesson to be noted is that for equations with nonlinear principle part one cannot tell if a given initial (Cauchy data) curve is suitable until very specific data on that curve are given.

5 COMPATIBILITY RELATIONS AND THE FINITE-DIFFERENCE METHOD OF CHARACTERISTICS

If k_1', k_2' are roots of (1.3.5), then on curve segments satisfying (1.3.7a, b) or (1.4.1a, b), which we have called characteristics, u_y and v_y are not uniquely determined even though u and v are known. We now ask under what conditions do u_y and v_y exist if u and v are known; i.e., can any combination of functions u and v be specified on a characteristic segment and still leave open the possibility of continuous extension across it (solution of the Cauchy problem on it), or must u and v somehow be compatible with the differential equation system when they are specified on a characteristic. To answer this question, we examine (1.3.4) and ask that

$$
\begin{vmatrix}
a_{12} - k_1' a_{11} & h_1 - a_{11}\left.\dfrac{du}{dx}\right|_1 - b_{11}\left.\dfrac{dv}{dx}\right|_1 \\[2mm]
a_{22} - k_1' a_{21} & h_2 - a_{21}\left.\dfrac{du}{dx}\right|_1 - b_{21}\left.\dfrac{dv}{dx}\right|_1
\end{vmatrix} = 0 \qquad (1.5.1a)
$$

and

$$
\begin{vmatrix}
a_{12} - k_2' a_{11} & h_1 - a_{11}\left.\dfrac{du}{dx}\right|_2 - b_{11}\left.\dfrac{dv}{dx}\right|_2 \\[2mm]
a_{22} - k_2' a_{21} & h_2 - a_{21}\left.\dfrac{du}{dx}\right|_2 - b_{21}\left.\dfrac{dv}{dx}\right|_2
\end{vmatrix} = 0. \qquad (1.5.1b)
$$

The student will find it sufficient here in this 2×2 case simply to think of solving (1.3.4) by Cramer's rule using determinants. If the denominator determinant is zero, then solutions are not unique, and in order for solutions to exist, the numerator determinants must be zero. These are (1.5.1a, b). On first glance, it might appear that two other conditions like (1.5.1a, b) need to be written with the right members of (1.3.4) replacing the first column of the determinant of the coefficient matrix, but an elementary principle of linear algebra assures us that the conditions so obtained will be equivalent to those of (1.5.1a) and (1.5.1b). The equivalence may be proved, of course, by direct manipulation but only at a cost of considerable labor. Notations $du/dx|_{1,2}$ and $dv/dx|_{1,2}$ in (1.5.1a, b) refer to derivatives on characteristics corresponding to roots k_1', k_2' of (1.3.5). We will call equations (1.5.1a, b) "compatibility relations." Only roots of (1.3.5) may appear in them because, otherwise, they are meaningless; by putting roots of (1.3.5) in (1.5.1a, b) we require that the numerator and denominator of Cramer's rule are zero together.

We now utilize (1.4.1a, b) and (1.5.1a, b) to devise a basic finite-difference procedure for numerical approximation to solutions of the Cauchy problem for hyperbolic type. Actually, it has a great deal more versatility, being adapatable to a great many boundary-value problems, but we present the finite-difference method of characteristics here for the assistance it can offer to the student in his efforts to understand the theory of characteristics.

Suppose $\gamma: (a, b) \to R^1$ and suppose u, v are given for every $x \in (a, b)$ and $y = \gamma(x)$. Let $\gamma \in C^1(a, b)$ and $\gamma' \neq k_{1,2}'$. We select a partition $x_1, x_2, \ldots, x_n \in (a, b)$, and let $x_1 = a, x_n = b$. For convenience, we set $x_i - x_{i-1} = h \in R^1$ for every $i = 1, \ldots, n$. Associated with any two points (x_1, y_1) and (x_2, y_2) (see the general cluster inset of Fig. 5) on the curve, where u and v are known,

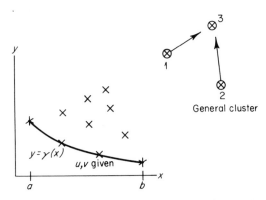

FIG. 5   The method of characteristics fills out a region of determination bounded on two sides by polygonal approximations of characteristic segments.

we specify how to approximate u and v at a third point (x_3, y_3) which lies approximately at the point of intersection of characteristics from points (x_1, y_1) and (x_2, y_2). From (1.4.1a, b) and (1.5.1a, b) we have

$$O(h) + \frac{y_3 - y_1}{x_3 - x_1} = k_1{}'(x_1, y_1, u_1, v_1) \tag{1.5.2a}$$

$$O(h) + \frac{y_3 - y_2}{x_3 - x_2} = k_2{}'(x_2, y_2, u_2, v_2) \tag{1.5.2b}$$

$$\begin{vmatrix} a_{12}^1 - k_1{}'a_{11}^1 & h_1 - a_{11}^1\left(\dfrac{u_3 - u_1}{x_3 - x_1} + O(h)\right) - b_{11}^1\left(\dfrac{v_3 - v_1}{x_2 - x_1} + O(h)\right) \\[2ex] a_{22}^1 - k_1{}'a_{21}^1 & h_2 - a_{21}^1\left(\dfrac{u_3 - u_1}{x_3 - x_1} + O(h)\right) - b_{22}^1\left(\dfrac{v_3 - v_1}{x_3 - x_1} + O(h)\right) \end{vmatrix} = 0 \tag{1.5.3a}$$

$$\begin{vmatrix} a_{12}^2 - k_2{}'a_{11} & h_1 - a_{11}^2\left(\dfrac{u_3 - u_2}{x_3 - x_2} + O(h)\right) - b_{11}^2\left(\dfrac{v_3 - v_2}{x_3 - x_2} + O(h)\right) \\[2ex] a_{22}^2 - k_2{}'a_{21}^2 & h_2 - a_{21}^2\left(\dfrac{u_3 - u_2}{x_3 - x_2} + O(h)\right) - b_{22}^2\left(\dfrac{v_3 - v_2}{x_3 - x_2} + O(h)\right) \end{vmatrix} = 0. \tag{1.5.3b}$$

Here a_{ij}^1, b_{ij}^1 stand for a_{ij}, b_{ij} evaluated at (x_1, y_1) while a_{ij}^2, b_{ij}^2 stand for a_{ij}, b_{ij} evaluated at (x_2, y_2); moreover, it is understood in (1.5.3a, b) that $k_1{}', k_2{}'$ are evaluated at $(x_1, y_1)(x_2, y_2)$, respectively. Deleting the indicated $O(h)$

terms, (1.5.2a, b) and (1.5.3a, b) give four linear equations in unknowns (x_3, y_3, u_3, v_3). The fact that γ is noncharacteristic assures a unique solution exists.

Solving of the equations can be reduced to successive solutions of 2×2 systems, a matter of some importance to convenience of computation. That is, since (1.5.2a, b) do not involve u_3 or v_3, they may be solved for (x_3, y_3), [geometrically the point of intersection of tangents to characteristics out of points (x_1, y_1) and (x_2, y_2)], and these values may be used in (1.5.3a, b) which can then be solved for (u_3, v_3). Of course, this can be repeated for all pairs of points in our implied partition of the noncharacteristic segment, giving rise then to a new "line" of points with one fewer point. On this new line the process can be repeated, dropping one point as before. If this process is continued, we eventually obtain a line containing a single point. An approximation to the solution has been obtained in a region, called a "characteristic triangle," which is bounded on two sides by polygonal approximations of characteristic segments, just as one would be led to suspect it would both from our analysis and picturization of the D'Alembert solution of the Cauchy problem as given in Section 1 and our exposition of the theory of characteristics. Of course, we have not proven existence and uniqueness of solutions of the Cauchy problem in general here, but one may now contemplate what has been presented for the wave equation in E^2 and for the theory of characteristics in order to see what is to be expected in general for hyperbolic type.

6 SYSTEMS LARGER THAN TWO BY TWO

We now return to (1.2.6) for discussion of those matters that require comment in the $n > 2$ case. Again, since we still consider only two independent variables (x, y) we can still use (1.3.3), only now in matrix (or vector) form,

$$U_x = dU/dx - k'U_y \tag{1.6.1}$$

and substitute this in (1.2.6). Then we have

$$(B - k'A)U_y = H - A \, dU/dx. \tag{1.6.2}$$

The characteristic equation now is

$$|B - k'A| = 0, \tag{1.6.3}$$

and the compatibility relations arise by asking that *on characteristics* the augmented matrix of (1.6.2)

$$(B - k'A; \; H - A \, dU/dx) \tag{1.6.4}$$

have the same rank as the coefficient matrix. Of course, there will be n linear factors of the polynomial (1.6.3). If some factor is repeated $k \leqslant n$ times, then the corresponding compatibility relation will be that the augmented matrix (1.6.4) have all subdeterminants of order $n-k$ equal to zero.

Classification is not really so clear any longer. If all roots are real and distinct, we call the system hyperbolic type; if all are real and equal, parabolic type; if n is even and all roots are complex conjugates, elliptic type. Otherwise, we refer to nonnormal type and many authors to mixed type, although we feel this last designation should be saved for equations which are of hyperbolic, parabolic, or elliptic type in different parts of a region.

7 FLOW AND TRANSMISSION LINE EQUATIONS

We seek to exercise our newly acquired skills on some "real," and preferably nonlinear, equations—ones that relate to some familiar, nontrivial physical phenomena and go beyond the elementary prototype examples of Section 3.

The (Euler) equations for inviscid, one-dimensional, time-dependent, compressible flow are

$$\rho_t + (\rho v)_x = 0 \qquad \text{(continuity)}$$
$$\rho v_t + \rho v v_x + p_x = 0 \qquad \text{(momentum)}$$

$(1.7.1)$

where, for some region $A \subset E^2$,

$$\rho, v, p: A \to R^1 \qquad \text{and} \qquad \rho, v, p \in C^1(A),$$

and the functional values represent density, velocity, and pressure, respectively. Independent variables x and t represent distance and time measured from some reference location and instant, respectively. A "continuum model" has been assumed in which density, velocity, and pressure values have been assigned to points in a region A. The "continuity equation" represents an expression of conservation of mass through volumes of arbitrary size, and the momentum equation expresses Newton's law for inertial force ($F = ma$). Such equations might be used for the analysis of gas flows in large cylindrical tubes, either in a number of devices now used in modern physics laboratories, or as an idealization of gas flows in gun barrels (Lagrange problem) or in a number of piston driven (or driving) devices.

Evidently the prescription (1.7.1) is incomplete; a relation between ρ and p, a so-called state equation "expressing the thermodynamics of the gas involved in the flow" must be given. Let $\Gamma: R^1 \to R^1$ and let $\Gamma \in C^1(R^1)$. Then we express a general state law as

$$p = \Gamma(\rho) \tag{1.7.2a}$$

or, using the chain rule (and our earlier announced convention for notations relating to composite functions), as

$$p_x = \Gamma'(\rho)\rho_x, \tag{1.7.2b}$$

and this addended to (1.7.1) gives the "real" system one wishes to study.

EXERCISE 7 Substitute (1.7.2b) into (1.7.1) and show that the resulting 2×2 system is of hyperbolic type. Find the formulas for characteristic directions (k_1', k_2') and find the compatibility equations.

Without argument, we will now state that the characteristics in the (x, t) plane may be regarded as traces of pressure waves (disturbances) moving down the tube.

EXERCISE 8 From the form of the characteristic directions obtained in Exercise 7, sketch the behavior of characteristics [in the (x, t) plane] in the presence of a compression wave moving down the tube and indicate that in this case the characteristics may tend to coalesce (form an envelope), thus forming a large gradient of pressure p_x. If this continues unchecked so that the pressure gradient gets larger and larger, a "shock" is said to be formed.

The transmission line equations are

$$
\begin{aligned}
-\partial i/\partial x &= g(v)v + c\,\partial v/\partial t \\
-\partial v/\partial x &= ri + l\,\partial i/\partial t
\end{aligned} \tag{1.7.3}
$$

where for a region $A \subset E^2$ we have $i, v \colon A \to R^1$ and $i, v \in C^1(A)$; $g \colon R^1 \to R^1$ or $g \in R^1$ (i.e., "or g is constant"); and $r, c, l \in R^1$. Of course, here functional values of i and v represent line current and voltage, respectively, while x and t again represent distance and time. The functional values of g—or the value of g if, as in the classical case, it is constant—represent leakage conductance. The fact that this term is included in the formulation makes these equations suitable for the treatment of marine and underground cables where, for excellent reasons that will develop much later in this text, some leakage is allowed. The coefficients r, c, and l are resistance, capacitance, and inductance per unit length of line.

EXERCISE 9 Show that (1.7.3) is of hyperbolic type and find the characteristics (not just their directions) and compatibility relations.

23

EXERCISE 10 Of course, (1.7.3) is nonlinear only because of the occurrence of the term $g(v)v$ (if g is not constant), and Exercise 9 brings out a significant difference in the nature of the nonlinearity in (1.7.3) and that in (1.7.1). Note this difference in specific terms.

EXERCISE 11 Let $i, v \in C^2(A)$. By differentiation and elimination of terms involving i, write a second-order equation in v which is equivalent to (1.7.3).

In deriving (1.7.3), it is assumed that the voltage can be treated as constant in any cross section of wire. Let us suppose that high frequencies are maintained on the wire in such a way that this assumption is not invalidated (i.e., that low enough frequencies are maintained so that wavelengths are long compared to the wire diameter). Then the induced voltage term in (1.7.3) heavily dominates the ohmic voltage drop ir, and the capacitive current term heavily dominates the line leakage term even if coefficients l and c are small, because for high frequencies the factors $\partial v/\partial t$ and $\partial i/\partial t$ are large in some parts of each wave. In this case, called a radiofrequency (rf), or inductance–capacitance (l–c) line, we may put $g(v) = r = 0$ for every $v \in R^1$.

EXERCISE 12 Using the results of Exercise 11 and those of Section 1, show that in rf circuits waves are transmitted at a constant velocity and without change of wave form.

EXERCISE 13 From the point of view of the theory of characteristics and classification of 2×2 first-order systems, discuss the difference in the nature of the perturbations invoked in deleting terms like the leakage gv or the ir drop in (1.7.3) and in dropping terms like the capacitive current $c\, \partial v/\partial t$ or induced voltage $l\, \partial i/\partial t$. The latter two are referred to as singular perturbations, the former as regular perturbations.

The (Euler) equations for inviscid, axially symmetric,[†] time-independent, compressible, and rotational flow are

$$(\rho u r)_r + (\rho v r)_z = 0 \qquad \text{(continuity)}$$

$$u u_r + v u_z + (1/\rho) p_r = 0 \qquad \text{(momentum)}$$

$$u v_r + v v_z + (1/\rho) p_z = 0 \qquad (1.7.4)$$

$$u s_r + v s_z = 0 \qquad \text{(entropy)}$$

$$p = \Gamma(\rho, s) \qquad \text{(state)}$$

---

† Depending only on z and on $r = (x^2 + y^2)^{1/2}$, not on x, y, and z independently.

where functional values of u and v represent velocity components in radial, r, and axial, z, directions, respectively. Of course, $s: A \to R^1$, $s \in C^1(A)$, and the functional values of s represent entropy. Entropy is a function used in thermodynamics in place of temperature which early in the history of that field was found "not to be a point function" since it is not an exact differential. The entropy differential is defined to be $ds = dQ/T$ where Q is heat content and T is temperature. The student who has not studied differentials in a form to make this statement meaningful may simply take the set of equations (1.7.4) as they are and is advised not to concern himself about this point at this time.

Let $K: R^1 \to R^1$ and let $\Gamma(\rho, s) = K(s)\rho^\gamma$ where γ is a real number. If s is constant, (1.7.4) is said to be isentropic, from which it follows[†] that curl v (v thought of as a function of three variables x, y, z) vanishes, so this case is also known as irrotational. The state law $p = K\rho^\gamma$ that also follows is called adiabatic, which refers to the fact that locally there is no net heat exchange.

Streamlines are curves that are always tangent to a velocity vector; here, such that $dz/dr = v/u$.

EXERCISE 14 Show that $(k' - v/u)^2$ is a factor of the characteristic equation for (1.7.4) whenever $\Gamma = K(s)\rho^\gamma$. Show that one of the compatibility relations corresponding to the streamlines viewed as characteristics gives that the entropy s is constant on any streamline.

Hint To obtain this last result note that a compatibility relation is an ordinary differential equation valid on a characteristic and get the result directly from the entropy equation. This will be much easier than to get it through the general theory (1.6.4).

The total energy, internal energy plus external work, is called enthalpy in thermodynamics and denoted by the symbol h. Like entropy, it is differentially defined:

$$dh = T \, ds + (1/\rho) \, dp. \tag{1.7.5a}$$

One may interpret (1.7.5a) as an equation of derivatives

$$dh/dt = T \, ds/dt + (1/\rho) \, dp/dt \tag{1.7.5b}$$

with respect to any parameter t.

[†] From manipulations of the momentum equations. This will be borne out by the work in Exercise 17.

EXERCISE 15 Show that the second compatibility relation corresponding to the streamlines as characteristics gives that

$$\frac{\gamma}{\gamma-1}\frac{p}{\rho} + \frac{1}{2}(u^2+v^2)$$

is constant on any streamline. This is called the Bernoulli law and expresses local conservation of energy for rotational flows.

Hint A combination of the momentum equations (and the entropy equation) gives the Bernoulli law as a compatibility equation easier than the general theory yields it.
Let $M = v/c$ where $c = (\partial\Gamma/\partial\rho)^{1/2} = (\gamma p/\rho)^{1/2}$; c is the velocity of sound, M the mach number.

EXERCISE 16 Find expressions for the other two characteristic directions (1.7.4) (i.e., other than $dz/dr = v/u$). Show that these directions are real and distinct if $M > 1$, real and equal if $M = 1$, and complex conjugates if $M < 1$.

We define a streamfunction for (1.7.4), $\psi : A \to R^1$ and $\psi \in C^1(A)$ up to a constant by

$$\psi_r = \rho v r, \qquad \psi_z = -\rho u r. \tag{1.7.6}$$

If we require that $\psi \in C^2(A)$, then the continuity equation reduces to $\psi_{rz} - \psi_{zr} = 0$ which is automatically satisfied [by requiring that ψ be related to u, v by the form (1.7.6) and that $\psi \in C^2(A)$]. The name "streamfunction" derives from the fact that on curves defined by $dz/dr = v/u$, we have

$$d\psi/dr = \psi_r + \psi_z\, dz/dr = \rho v r - \rho u r(v/u) = 0$$

so that ψ is constant on any streamline. From the result of Exercise 14, we may put $s = s(\psi)$ and write

$$p = A(\psi)\rho^\gamma.$$

In Exercise 15 we found that one combination of the momentum equations in (1.7.4) gave the Bernoulli law.

EXERCISE 17 Show that another combination of the momentum equations gives a streamfunction equation

$$A\psi_{rr} + 2B\psi_{rz} + c\psi_{zz} = D \tag{1.7.7}$$

where

$$A = (1 - u^2/c^2), \qquad B = -(uv/c^2), \qquad C = (1 - v^2/c^2)$$

and for the isentropic case ($s = $ constant) $D = \rho v$.

Hint The isentropic case is the easiest and should be done first.

Of course, (1.7.7) is to be taken together with the Bernoulli law, the "algebraic" (i.e., not differential) equation derived in Exercise 15. Also, it must be accompanied by the state law, another "algebraic" equation, but these equations do not affect considerations of classification, so for this purpose we may direct our attention to the single second-order equation (1.7.7).

EXERCISE 18 Using the result (1.3.9) for the general quasi-linear second-order equation (1.2.2), show that (1.7.7) is of hyperbolic, parabolic, or elliptic type as M is $>$, $=$, or <1.

It is from the considerations of the streamfunction equation (1.7.7) as in Exercise 18 that one often hears supersonic, sonic, or subsonic flows identified with hyperbolic, parabolic, or elliptic type. The student will want to compare the results of Exercise 16 on the Euler equations with those of Exercise 18 on the corresponding streamfunction equation. It should be carefully noted that these results are for the very special cases of time-independent flows; the results of Exercises 16 and 18 should be compared with those of Exercise 7 where it can be seen even from a simple model that time-dependent inviscid flows are always of hyperbolic type! The addition of viscous effects changes this picture as we will now see.

The velocity function v for incompressible ($\partial \Gamma / \partial \rho = 0$), axially symmetric, time-dependent, viscous flow through small cylindrical tubes with rigid walls satisfies the equation

$$v_{rr} + rv_r - (1/(\mu/\rho))v_t = -f(t)/\mu \tag{1.7.8}$$

where $f: (t_0, \infty) \to R^1$ and values $f(t)$ represent pressure gradients at time t which we regard as given (i.e., measured). Here, μ is the viscosity coefficient, ρ is density (constant), and r is the radial variable. Because of the incompressibility of the fluid and the rigidity of the walls—i.e., if these are, indeed, strictly valid—the axial variable z does not occur.

EXERCISE 19 Show that Eq. (1.7.8) is of parabolic type with characteristics $r = $ constant.

The origins of both Eqs. (1.7.3) and (1.7.8) from physical principles will eventually be described here and extensive treatments will be given of them. The equations for inviscid flow will also come in for some further discussion in the context of efforts at numerical approximation. Usually equations with nonlinear principle part do not lend themselves readily to theoretical treatment. However, in a use of the finite-difference method of characteristics a computing machine cannot tell, so to speak, that the equations are not linear; each step forward is simply taken from known positions using local linearizations. The following exercise emphasizes the importance of the complications brought on by an equation having a nonlinear principle part.

EXERCISE 20 For either (1.7.1) or (1.7.4) we wish to specify the unknown flow variables on a curve segment, $x = \gamma(r)$ or $z = \gamma(r)$, thus conceiving initial-value or Cauchy problems. Discuss the question of validity for such problems in the light of the theory of characteristics as discussed in Section 4.

2

Various Boundary-Value Problems for the Homogeneous Wave Equation in E^2

1 THE CAUCHY OR INITIAL-VALUE PROBLEM

The Cauchy or initial-value problem has been thoroughly discussed in the beginning considerations of Chapter 1. The boundary-value problem (1.1.1), (1.1.2) pictured in Fig. 1 has the unique solution (1.1.12), known as the D'Alembert solution. The qualitative properties of the solution are pictured in Fig. 2 where it is emphasized that characteristics $x \pm y = $ constant form natural boundaries for the region of determination, a fact which helps to illuminate the more general theory of characteristics.

2 THE CHARACTERISTIC BOUNDARY-VALUE PROBLEM

With $a < 0 < b$, consider the homogeneous wave equation (1.1.1) and the boundary conditions

$$u(x, x) = f(x) \quad \text{and} \quad u(x, -x) = g(x) \tag{2.2.1}$$

for $g: (a, 0) \to R^1$ and $f: (0, b) \to R^1$. We are to ask once again in what region R does the data f, g determine a unique solution and what properties of f and g must we require in order that a solution exist in R. Moreover, we must again decide in unequivocal terms what it is that we will call a solution, sticking for now with the simplest possible interpretation of this concept.

Once again, all solutions are of the form

$$u(x,y) = F(x+y) + G(x-y) \qquad (2.2.2)$$

where F, G must now be determined in terms of f, g given in (2.2.1). We have immediately that

$$f(x) = F(2x) + G(0) \qquad (2.2.3)$$

$$g(x) = F(0) + G(2x) \qquad (2.2.4)$$

or

$$F(x) = f(x/2) - G(0)$$

$$G(x) = g(x/2) - F(0)$$

or, using (2.2.2),

$$u(x,y) = f(\tfrac{1}{2}(x+y)) + g(\tfrac{1}{2}(x-y)) - (F(0)+G(0)).$$

But from (2.2.2), (2.2.3), and (2.2.4)

$$u(0,0) = f(0) = g(0) = F(0) + G(0), \qquad (2.2.5)$$

and this reveals another condition,

$$f(0) = g(0), \qquad (2.2.6)$$

that must be added to (2.2.1) in order to obtain a solution in the form (2.2.2). Then we have

$$u(x,y) = f(\tfrac{1}{2}(x+y)) + g(\tfrac{1}{2}(x-y)) - f(0). \qquad (2.2.7)$$

A quick examination shows that (2.2.7) satisfies (2.2.1) whenever (2.2.6) holds, and it is of the form (2.2.2) so that it must satisfy the differential equation, but where is the expression (2.2.7) valid? Once again we ask what is the solution at a point (x_0, y_0) (see Fig. 6), and once again we draw the two characteristics

$$x - y = x_0 - y_0 \qquad \text{and} \qquad x + y = x_0 + y_0 \qquad (2.2.8)$$

through this point. The first intersects the boundary $x+y = 0$ at the point $\frac{1}{2}(x_0 - y_0)$, which is where we wish to pick up the value of g to be used in (2.2.7), and the second intersects the boundary $x - y = 0$ at the point $\frac{1}{2}(x_0 + y_0)$ which is where we wish to pick up the value of f to be used in (2.2.7). Thus the solution is given by (2.2.7) in the rectangle R bounded by $x - y = 0$, $x + y = 0$, the given boundaries, and $x - y = 2a$, $x + y = 2b$, the "natural"

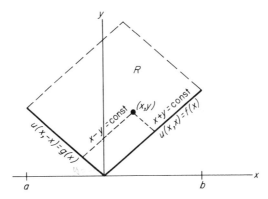

FIG. 6   The characteristic boundary-value problem.

boundaries. This will be the region of determination, for it is evident in the above that $g(x) = 0$ for $x \in (a, 0)$ and $f(x) = 0$ for $x \in (0, b)$ implies $u(x, y) \equiv 0$ for $(x, y) \in R$. The student should carefully notice that it was said in the last chapter not that data should never be given on characteristics, but only that Cauchy data on a characteristic segment do not uniquely determine a solution. In fact, this problem, where data are given on two characteristics, serves further to elucidate those principles introduced in the last chapter. When one piece of data is given on a characteristic, if the differential equation is assumed valid there, this determines the other piece of data at least up to a constant, and unique determination requires the specification of another piece of data on an intersecting line, here a characteristic of the other family.

We have tacitly called a function u a solution if

(i) $u \in C^2(R)$;

(ii) $u \in C(R \cup \delta R)$;

(iii) u satisfies $u_{xx} - u_{yy} = 0$ in R; and

(iv) u satisfies (2.2.1), (2.2.6).

Though the Cauchy problem is by far the more popular problem for hyperbolic type among mathematicians, we will find that in one of the important general classical theories of hyperbolic type—the Riemann method—the characteristic boundary-value problem is made to play a basic role, providing a resolvent function to be used in representing solutions to other boundary-value problems.

3 THE MIXED BOUNDARY-VALUE PROBLEM

The line intersecting a characteristic where data are given would not seem to have to be a characteristic. Only a feeling for symmetry has caused us first to look at the case of data on two intersecting characteristics rather than on one characteristic and one noncharacteristic. Of course, the latter case can be regarded now as a mixed problem—a combination of a characteristic boundary-value problem where the specification of data on one of the two characteristics has been ignored and a Cauchy problem where specification of the derivative data has been ignored. In this sense, the following problem is a combination of those studied in Sections 1 and 2. Once again the homogeneous wave equation in E^2 is to be satisfied with

$$u(x,0) = f(x) \qquad \text{and} \qquad u(x,x) = g(x) \qquad (2.3.1)$$

where $f: (0,a) \to R^1$, $g: (0,a) \to R^1$, and again we are to ask where and under what conditions on f and g is a unique solution determined. Once again, we simply evaluate F and G in the general solution

$$u(x,y) = F(x+y) + G(x-y)$$

of the differential equation, whence

$$f(x) = F(x) + G(x)$$
$$g(x/2) = F(x) + G(0)$$

or

$$G(x) = f(x) - g(x/2) + G(0)$$
$$F(x) = g(x/2) - G(0)$$

or

$$u(x,y) = g(\tfrac{1}{2}(x+y)) - g(\tfrac{1}{2}(x-y)) + f(x-y) \qquad (2.3.2)$$

where the value $G(0)$ has been added and subtracted out. One sees immediately that the condition

$$f(0) = g(0) \qquad (2.3.3)$$

must be added and then (2.3.1) is satisfied by (2.3.2). Also, one sees that (2.3.2) is in the form of a solution of the differential equation, and the region of determination is the region in Fig. 7 bounded by the two characteristic segments through points $(a,0)$ and (a,a) (which boundary segments were not specified) as well as the given segments on $y = 0$ and $y = x$. The solution exists in this region if $f \in C^2(0,a)$ and $g \in C^2(0,a)$. That the solution is then unique in this region is again immediate.

32

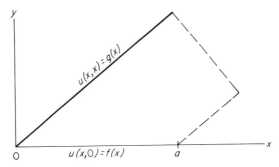

FIG. 7 The mixed boundary-value problem. Primitive data are given on a characteristic and a noncharacteristic.

4 THE GOURSAT PROBLEM

We have begun now to sample the specifications of various boundary-value problems and our curiosity should be aroused about the range of possibilities. Representative cases have been carefully chosen for the purpose of generating this curiosity. The impression from the above, however, that it will always be trivial to answer questions concerning existence and uniqueness is certainly invalid. The two problems just taken up are surely among the few for which it is true. It is not, for example, true for the problem that should occur next to one who has become curious about boundary-value problems.

It should now occur to one to ask if there is a unique solution to the homogeneous wave equation determined by specification of data on two intersecting noncharacteristics strictly contained in an angle between characteristics. In accordance we investigate the specification (see Fig. 8)

$$u(x,0) = f(x) \quad \text{and} \quad u(x,\beta x) = g(x) \tag{2.4.1}$$

where

$$f: (0,a) \to R^1, \quad g: (0,a) \to R^1, \quad f(0) = g(0) \tag{2.4.2}$$

and

$$0 > \beta > 1. \tag{2.4.3}$$

For $\beta = 0$, the problem degenerates, and for $\beta = 1$ it becomes the mixed problem of Section 3.

Once again we have from the general solution

$$u(x,y) = F(x+y) + G(x-y),$$

33

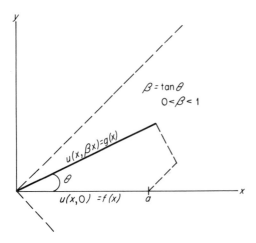

FIG. 8 The Goursat problem. Primitive data are specified on two noncharacteristics in the angle between characteristics.

using the boundary conditions (2.4.1), that

$$f(x) = F(x) + G(x) \tag{2.4.4}$$

and

$$g(x) = F((1+\beta)x) + G((1-\beta)x), \tag{2.4.5}$$

but this set of function equations is not so trivial to solve for F and G as the previous ones have been. Toward solving them, or really toward finding conditions on f and g so that (2.4.4), (2.4.5) are uniquely solvable for F and G, we rewrite them as one equation in G. That is, from (2.4.4) we put

$$F(x) = f(x) - G(x), \tag{2.4.6a}$$

or

$$F((1+\beta)x) = f((1+\beta)x) - G((1+\beta)x), \tag{2.4.6b}$$

into (2.4.5) to obtain

$$g(x) - f((1+\beta)x) = -G((1+\beta)x) + G((1-\beta)x).$$

Here we let

$$x = \frac{X}{1+\beta}, \quad 0 < \alpha = \frac{1-\beta}{1+\beta} < 1, \quad p(X) = g\left(\frac{X}{1+\beta}\right) - f(X) \tag{2.4.7}$$

so that this becomes

$$p(X) = -G(X) + G(\alpha X). \tag{2.4.8}$$

34

This is the function equation we must solve for G subject to conditions on p, a given combination of functions f and g. From (2.4.8)

$$p(\alpha X) = -G(\alpha X) + G(\alpha^2 X),$$

$$p(\alpha^2 X) = -G(\alpha^2 X) + G(\alpha^3 X),$$

$$\vdots$$

$$p(\alpha^n X) = -G(\alpha^n X) + G(\alpha^{n+1} X),$$

and because of cancellations on successive rows, adding these n expressions to that of (2.4.8) provides the result

$$\sum_{i=0}^{n} p(\alpha^i X) = G(\alpha^{n+1} X) - G(X).$$

Since G must at least be continuous and $0 < \alpha < 1$,

$$G(X) = G(0) - \sum_{i=0}^{\infty} p(\alpha^i X) \qquad (2.4.9)$$

if, indeed, the limit indicated by writing the infinite sum on the right does exist. In that case, from (2.4.6a)

$$F(X) = f(X) + \sum_{i=0}^{\infty} p(\alpha^i X) - G(0), \qquad (2.4.10)$$

and the solution to our problem is simply

$$u(x, y) = f(x+y) + \sum_{i=0}^{\infty} p(\alpha^i (x+y)) - \sum_{i=0}^{\infty} p(\alpha^i (x-y)), \qquad (2.4.11)$$

where the quantities $G(0)$ and $-G(0)$ have summed to zero.

Thus our problem is solved if conditions are given on f and g so that

$$\sum_{i=0}^{\infty} p(\alpha^i X)$$

converges where p is defined by (2.4.7).

EXERCISE 1 Specify sufficient conditions on f and g so that $\sum_{i=0}^{\infty} p(\alpha^i X)$ is dominated by a convergent geometric series of positive terms.

Hint Notice that $p(0) = 0$.

EXERCISE 2 Show that (2.4.11) reduces to (2.3.2) whenever $\beta = 1$.

5 THE VIBRATING STRING PROBLEM

For the continuum model, the vibrating string problem is a boundary-value problem for the homogeneous wave equation in E^2 where it is specified that

$$u(x,0) = f(x), \qquad u_y(x,0) = g(x) \tag{2.5.1}$$

and

$$u(0,y) = h_0(y), \qquad u(l,y) = h_l(y) \tag{2.5.2}$$

where

$$f, g \colon (0, l) \to R^1, \qquad h_0, h_l \colon (0, Y) \to R^1.$$

Although historically this was perhaps the first boundary-value problem to be considered, the questions of existence and uniqueness can now be answered by composition of this problem from the simpler ones—characteristic, initial value, and mixed—treated in Sections 1, 2, and 3. Presumably, all the problems for the homogeneous wave equation in E^2 can be readily composed from the four basic problems given above or from some straightforward transformations of them. Here a solution is seen to be determined in the characteristic triangle (region 1) above the base (see Fig. 9) by the D'Alembert solution of the Cauchy problem whenever we require that

$$f \in C^2(0, l), \qquad g \in C^1(0, l).$$

This solution in turn determines values of u on the characteristics $y = x$ and

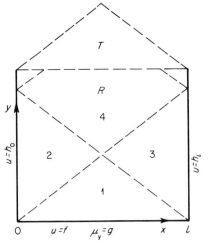

FIG. 9   The vibrating string problem.

$y = -x + l$ (which bound region 1) up to the point $(l/2, l/2)$ and together with boundary values h_0, h_l on $x = 0$, $x = l$ determines solutions in regions 2 and 3 by solution of the mixed problem. Of course, we require that

$$h_0(0) = f(0), \qquad h_l(l) = f(l), \qquad \text{and} \qquad h_0, h_l \in C^2(0, Y).$$

But now we can obtain u on the characteristic segments lying just under the region 4 where solutions can be determined from the characteristic boundary-value problem. In this manner the entire rectangle R can be filled out in which the solution is uniquely determined, and even a unique continuation can be made to cover the triangle T attached to the top of this rectangle.

EXERCISE 3 Using $g = h_0 = h_l = 0$, write the solution of the vibrating string problem in regions 1, 2, 3, and 4 of Fig. 9 by composing the solution from solutions given in Sections 1, 2, and 3.

The usual way of generating solutions is not that given in Exercise 3, this technique being presented only to elucidate the handling of existence and uniqueness by composition from the four problems above which we will now regard as basic. If numbers are needed for functional values of the solution, one will usually be well advised to use numerical methods. This is especially advisable if h_0 and h_l are not to be put equal to zero. The method of characteristics outlined in Chapter 1 could be utilized where the sides are used like characteristics and the functions h_0, h_l are used like compatibility conditions to obtain new points on $x = 0$ and $x = l$. This will yield an n-vector, x^{k+1}, of values on the $(k + 1)$st line, in terms of the n-vector, x^k, of values on the kth line. Since the wave equation is linear and with constant coefficients, a square matrix M and column matrix h of real numbers can be found so that this relation is expressed in the form

$$x^{k+1} = Mx^k + h. \tag{2.5.3}$$

EXERCISE 4 Find M and h for the vibrating string problem.

Of course, since the initial values are given, (2.5.3) allows computation[†] of an approximate solution to the vibrating string problem, successively line by line. The question of what mesh sizes are required for the approximation to be a good one when the components of x^k and x^{k+1} are taken to be any real

---

† The act of computing, however, should probably be done in component, not matrix, form.

numbers generated by (2.5.3) is perhaps significant, but it rarely causes any practical difficulty. Another difficulty is much less obvious but far more important to be aware of from a practical standpoint. In any computation terminating decimals must be used. Thus the components of x^{k+1} and x^k recorded in any computation are terminated at a given number of (decimal) digits. This results, of course, in an error vector δ^0 even on the initial line, and any line may be considered the initial line. Assuming that the side data and other parameters going to make up components of M and h are represented exactly, we put

$$x^{k+1} + \delta^{k+1} = M(x^k + \delta^k) + h \tag{2.5.3a}$$

where δ^k is the error thus made on line k, due to round-off on the initial line, and δ^{k+1} is the error on line $k+1$ computed exactly from (2.5.3). We wish to assess whether or not such an error vector continues to grow in magnitude assuming that all successive lines of values obtained from it are computed and recorded exactly. *Since the problem is linear, there is no inconsistency in thus treating effects of errors one line at a time.*

Subtracting (2.5.3a) from (2.5.3),

$$\delta^{k+1} = M\delta^k = M^2\delta^{k-1} = \cdots = M^{\underline{k+1}}\delta^0, \tag{2.5.4}†$$

or, as we now prefer to see it,

$$\delta^{k+1} = M(M^{\underline{k}}\delta^0). \tag{2.5.4a}$$

For large k, the last indicated multiplication by M (on the left) may as well be a real symmetric matrix, similar to the diagonal matrix Λ of eigenvalues of M listed in nondecreasing order. The reason is that, in general, M is actually similar to such a matrix with possibly ones appearing opposite repeated diagonal elements, if such repetitions occur (Jordan form). We have

$$M \sim \Lambda + \Psi,$$

where Ψ commutes with Λ and $\Psi^p = 0$ for some $p \geqslant n$. Using then the binomial theorem, we find that for large k,

$$M^{\underline{k}} \sim (\Lambda + \Psi)^{\underline{k}}, \tag{2.5.5}$$

behaves like

$$\Lambda^{\underline{k}}. \tag{2.5.6}$$

---

† The superscript k on δ indicates the kth line while the superscript $\underline{k}$ on M indicates M is to be multiplied by itself k times (a power).

38

The marching scheme (2.5.3) will be considered (numerically) asymptotically stable if for large k,

$$\max_i |\delta_i^{k+1}| \leqslant \max_i |\delta_i^k|, \qquad (2.5.7)$$

and this will be true, according to (2.5.4) and (2.5.6), if

$$\max_i |\lambda_i| \leqslant 1, \qquad (2.5.8)$$

where λ_i, $i = 1, \ldots, n$, are the eigenvalues (possibly complex) of M. It is not necessary here that δ_i^{k+1} goes to zero with k and, therefore, not necessary that (2.5.8) be a strict inequality. It turns out that the method of characteristics is always stable.

EXERCISE 5 From (2.5.8), using the results of Exercise 4, show that the method of characteristics for the vibrating string problem is stable.

EXERCISE 6 Writing the wave equation as a first-order system, generate a marching scheme of the form (2.5.3) for the vibrating string problem by replacing the system directly with finite differences on a cluster of three points, one above and half way between the other two. Use an "artifiicial" point as one where the values are the average of the two lower points in forming the replacement for vertical derivatives, and make use of a special procedure for side or endpoints of a line.

EXERCISE 7 Show that the direct finite difference replacement for the vibrating string problem is stable when the upper point falls on or within the characteristic (isosceles, 45°) triangle based on the other two.

For the physical case of a vibrating string, y is time, x is distance from one end, and u is displacement. Thus if the ends are fixed, we do indeed have $h_0 = h_l = 0$. Moreover, with $h_0 = h_l = 0$, the solution with $u(x,0) = f(x)$, $u_y(x,0) = g(x)$ is the sum of solutions with $u(x,0) = f(x)$, $u_y(x,0) = 0$ and $u(x,0) = 0$, $u_y(x,0) = g(x)$. Thus the problem in Exercise 3 is not so artificial as it may at first seem to be. The classical approach to this problem is by way of trigonometric series. For example, suppose that f has a convergent sine series,

$$f(x) = \sum_{n=0}^{\infty} a_n \sin nx. \qquad (2.5.9)$$

39

Then one asks if there is a sum of terms satisfying the condition $u(x, 0) = f(x)$, $u_y(x, 0) = 0$, $u(0, y) = u(l, y) = 0$ so that each term is of the form

$$U(x, y) = \varphi(x)\psi(y)$$

and satisfies the homogeneous wave equation. This is then called the Fourier series solution; the method is called separation of variables.

EXERCISE 8 Formally deduce the Fourier series that would solve the problem in Exercise 3 if (2.5.9) holds.

This manipulation is called formal because we have not tackled the question of convergence. Actually, (2.5.9) (or a modification allowing cosine terms) is valid under extremely general circumstances, and under such circumstances the series almost always converges, at least in mean square. This is the well-appreciated advantage of the method. Since engineers find it easy to generate sine waves electrically, they also find it easy to compose electrical analogues of solutions given in this form. This mathematical technique requires considerable attention, and though often taught at an elementary level, requires very advanced concepts for a sound and thorough treatment. We do not treat it further here, assuming the student either will have already studied it in a form as good as we could present here or that he will later meet it in a better form, perhaps in a course in real variables. This technique, and its very substantial ancillary developments, simply are not our concern in this text. We shall use it when this is illuminating or allude to it at will, but we shall not study it here.

6 UNIQUENESS OF THE VIBRATING STRING PROBLEM

An alternative proof of uniqueness of the vibrating string problem will be interesting especially inasmuch as it will be found to introduce a class of techniques that is quite successful in problems related to both elliptic and parabolic as well as hyperbolic type. Let R be the rectangle shown in Fig. 9 where now the uniqueness problem is to be considered in R, so we have that for $y = 0$, $u = u_y = 0$, and $u = 0$ whenever $x = 0$ or l. Let Y' be any real number such that $0 < Y' \leqslant Y$ and denote by R' the portion of R lying beneath the line $y = Y'$ (see Fig. 10). We will prove the uniqueness of regular solutions in R defined in the following way:

A function $u: R \cup \delta R \to R^1$ is said to be a regular solution of the vibrating string problem in R if

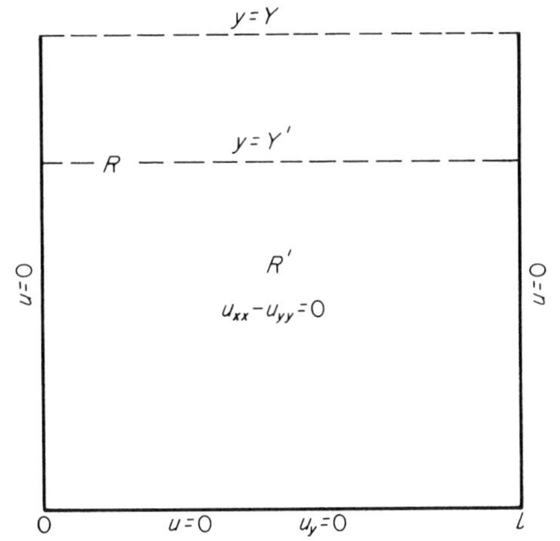

FIG. 10 If the wave equation is satisfied in R, it is satisfied in any subrectangle R'.

(i) $u \in C^2(R)$,
(ii) $u_{xx} - u_{yy} = 0$ in R,
(iii) $u \in C^1(R \cup \delta R)$, and
(iv) the boundary conditions (2.5.1) and (2.5.2) are satisfied.

The reason for needing C^1 and not just C in (iii) for this proof will be evident in a moment. It may be noted that for the uniqueness theorem obtained by composition from the basic problem, only C was necessary in (iii).

Since the wave equation must be satisfied in any region R', we have

$$0 = \int_{R'} u_y(u_{xx} - u_{yy})\, dR = \int_{R'} [(u_y u_x)_x - \tfrac{1}{2}(u_x^2)_y - \tfrac{1}{2}(u_y^2)_y]\, dR$$

$$= \int_0^{Y'} dy \int_0^l (u_x u_y)_x\, dx - \tfrac{1}{2} \int_0^l dx \int_0^{Y'} (u_x^2 + u_y^2)_y\, dy$$

$$= \int_0^{Y'} u_x u_y \big|_{x=0}^{x=l} - \tfrac{1}{2} \int_0^l [u_x^2(x,Y') + u_y^2(x,Y')]\, dx, \qquad (2.6.1)$$

where in the last integral we have made use of the fact that both u_y and u_x vanish on the line $y = 0$. That u_x vanishes there follows from the fact that u does. The entire integrand of the first integral vanishes because u_x is bounded on $R \cup \delta R$ [see (iii) above], and because $u = 0$ on $x = 0,\ l$ implies $u_y = 0$ on $x = 0, l$. Then from (2.6.1), for any $Y' \in (0,Y)$

$$u_x(x,Y') = u_y(x,Y') = 0$$

for every $x \in (0,l)$; i.e., $u_x = u_y = 0$ in R, or, by continuity, even in $R \cup \delta R$. But then u is constant in $R \cup \delta R$, and since it is zero on one part of δR we conclude that $u = 0$ on $R \cup \delta R$.

41

7 THE DIRICHLET PROBLEM FOR THE WAVE EQUATION?

As we continue to develop our curiosity about the range of possibilities for boundary-value problems, we will now begin to wonder about a possibility for modifying the vibrating string problem. For example, could one hope to obtain a good problem by replacing the derivative data u_y given in the vibrating string problem on the line $y = 0$ by primitive data u given on the line $y = Y$. This would be a problem in which $u_{xx} - u_{yy} = 0$ on the interior of a rectangle and where u would be specified over the entire closed boundary surrounding it (see Fig. 11). Of course, we would also ask for a continuity condition to be satisfied on the closure of a region. It will be found that this type of problem is one of the basic ones for the Laplace equation (and other equations of elliptic type), and it is known as a Dirichlet problem. One could put the question here, why is it not declared basic for hyperbolic type as well?

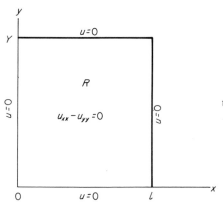

FIG. 11   The uniqueness question for the Dirichlet problem for the wave equation in E^2.

Let us consider the uniqueness problem. If we can generate nontrivial solutions for the homogeneous boundary-value problem ($u_{xx} - u_{yy} = 0$ in $R, u = 0$ in δR) by any device whatsoever, it is clear that the solution is not unique. In fact, such solutions of the homogeneous boundary-value problem could simply be added to any solution proposed for a nonhomogeneous problem, thus generating more solutions for the given problem. We seek such solutions by the special technique of separation of variables; i.e., we let

$$u = \varphi(x)\psi(y),$$

so that from the wave equation

$$\frac{\varphi''(x)}{\varphi(x)} = \frac{\psi''(y)}{\psi(y)},$$

and both must be equal to a constant. We choose the constant $-k^2$ because if we used a positive constant, it is clear that our homogeneous boundary conditions could not be satisfied. We thus have that

$$\varphi'' + k^2 \varphi = 0 \quad \text{and} \quad \psi'' + k^2 \psi = 0.$$

Then

$$\varphi(x) = A \sin kx + B \cos kx,$$

but from $\varphi(0) = 0$, we have $B = 0$. Then from $\varphi(l) = 0$, we have

$$0 = A \sin(kl),$$

which is satisfied (even though $A \neq 0$) if

$$kl = n\pi \quad \text{or} \quad k = n\pi/l, \quad n = 1, 2, \dots.$$

Similarly, the homogeneous boundary conditions at $y = 0$ and $y = Y$ will be satisfied if

$$kY = m\pi \quad \text{or} \quad k = m\pi/Y, \quad m = 1, 2, \dots,$$

and still not give trivial values for $\psi(y)$. We then would have the class of nontrivial solutions

$$u(x, y) = A_n \sin(n\pi x/l) B_m \sin(m\pi y/Y).$$

However, it is clear in the above that for this to be valid we must have that

$$k = n\pi/l = m\pi/Y \quad \text{or} \quad Y/l = m/n.$$

But m and n are integers. We have shown that the Dirichlet problem is not unique on a rectangle if the ratio of its height to length of base is rational!

One can see in the above that no nontrivial separable solution of the homogeneous boundary-value problem can be found if the ratio Y to l is irrational. Certainly the reader will readily agree without proof here that the symmetry of the boundary, boundary conditions, and differential equation are such for this problem that any solutions would have to be separable or have to be a sum of separable solutions. With this agreement we now have seen that *the Dirichlet problem for the wave equation on a rectangle is unique if the ratio of the length of its sides are irrational; otherwise, the solution is not unique!*

This curious fact brings up a bizarre instability feature. By passing from an irrational to rational length ratio one can introduce an infinitude of nontrivial solutions that could be added to any proposed solution corresponding to specified boundary data. For this reason, it would seem that Dirichlet problems for the wave equation (at least on a rectangle) could never be useful in describing phenomena of nature. Many examples of boundary-value problems are not considered useful because some sort of stability difficulties arise in considering applications of partial differential equations. This particular stability difficulty will be rare. For some others, notably the Cauchy problem for the Laplace equation, we will find that sufficient care in formulating a problem will transform it into a useful one.

3

Various Boundary-Value Problems for the Laplace Equation in E^2

1 THE DIRICHLET PROBLEM

One who has come to understand how bizarre the Dirichlet problem is for the wave equation with respect to uniqueness[†] may find it a riddle that this problem is considered basic for the Laplace equation, since this equation differs from the wave equation in one sign only. However, such differences with respect to existence and uniqueness when making a change in sign are quite common in mathematics, so common in fact, that the reader may not have noted it before. An entirely analogous algebraic problem requires that we find real roots for y as a function of x when

$$\frac{x^2}{a^2} + \frac{y^2}{b^2} = 1 \quad \text{and} \quad \frac{x^2}{a^2} - \frac{y^2}{b^2} = 1.$$

Of course, the first problem has two solutions $x = \pm a$, while the second has none (see Fig. 12).

We have learned by now that it is well to start out by defining what we will mean by a solution. To save repetition later we will do this now, in general, in

---

[†] Obviously, also, for existence of separable solutions.

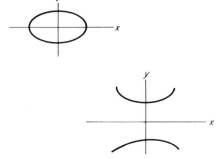

FIG. 12 The function of x derived from the equation of an ellipse has two real roots; for a similar hyperbola, it has none.

n-dimensional space, but, sticking to the spirit of the outline promised, we will undertake in this chapter only problems in two dimensions. In fact, only a very restricted class of these problems will be presented here. Further treatments of the Laplace equation will be given in Chapter 10 on potential theory.

Let A be a region (open, connected set) in E^n and let δA be its topological boundary (the set of limit or "cluster" points not in A). A function u is said to be a solution of the Dirichlet problem for the Laplace equation on A with respect to data f if

(i) $u \in C^2(A)$,
(ii) $\sum_{i=1}^{n} u_{x_i x_i} = 0$ in A,
(iii) $u \in C^1(A \cup \delta A)$, and
(iv) $u = f$ in δA.

Here it is understood that a function $f: \delta A \to R^1$ is given and that an x in A can be written as $(x_1, \ldots, x_n)$ (see Fig. 13).

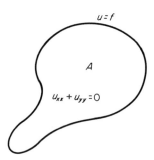

FIG. 13   The Dirichlet problem for the Laplace equation in E^2.

2 RELATION TO ANALYTIC FUNCTIONS OF A COMPLEX VARIABLE

The theory of the Laplace equation in two real variables x and y is virtually an image of the theory of analytic functions of one complex variable $z = x + iy$. A complex-valued function of a complex variable is said to be analytic at a point z_0 of the complex plane if it is differentiable in some neighborhood of z_0. For such a function to be differentiable at z_0 the limit

$$\lim_{h \to 0} \frac{f(z_0 + h) - f(z_0)}{h}$$

must exist and be the same value, whatever pathway $z_0 + h$ describes, as h (a complex number) decreases to zero. In particular, we must have that

$$\partial f / \partial x = -i \, \partial f / \partial y$$

since these represent derivatives along straight paths parallel to the x and y axes. Putting

$$f(z) = u(x, y) + iv(x, y)$$

and carrying out this differentiation gives the first-order partial differential equation system called the Cauchy–Riemann equations which were already mentioned in connection with Exercise 2 of Chapter 1; then differentiating this system appropriately shows that the real and imaginary parts of an analytic function of a complex variable satisfy the Laplace equation.

EXERCISE 1 Derive the Cauchy–Riemann equations and show how they lead to the fact that the real and imaginary parts of an analytic function of a complex variable satisfy the Laplace equation.

We now consider a region D of the complex plane, the topological boundary C of which is a simple closed rectifiable curve. Let f be analytic in D (i.e., at all points of D) and continuous in $C \cup D$. The Cauchy integral theorem (strong form) states that

$$\oint_C f(\zeta) \, d\zeta = 0, \tag{3.2.1}$$

and the Cauchy integral formula says that

$$f(z) = \frac{1}{2\pi i} \oint_C \frac{f(\zeta)}{\zeta - z} \, d\zeta, \tag{3.2.2}$$

where it is to be noted that though $f(\zeta)$ is analytic in D and continuous on

47

$D \cup C$, the function $f(\zeta)/(\zeta - z)$ is singular at the boundary point $\zeta = z$ (see Fig. 14). In fact, (3.2.2) is a singular integral. The important thing here is to note that (3.2.2) gives the value of f at any interior point z in terms of the values of f at its boundary points ζ. In this sense it is a solution of a Dirichlet problem, especially since, as we have noted, the real and imaginary parts satisfy the Laplace equation.

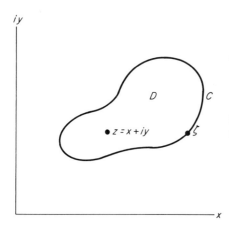

FIG. 14 The Cauchy integral formula is essentially the solution of a Dirichlet problem.

Of course, we can speak of the Laplace equation $\sum_{i=1}^{n} u_{x_i x_i} = 0$ and the Dirichlet problem in any number of dimensions. It is this higher-dimensional feature which causes us to regard the study of the Laplace equation as an extension of the theory of analytic functions of a complex variable. Without an extension in dimension, we can so regard the study of equations of elliptic type. Other types, especially hyperbolic type, offer more a contrast than an extension since, as we have seen, it is not always possible in other types to make power series (analytic) extensions across certain boundaries. We will find in our study of the Cauchy–Kovalevski theorem that for equations of elliptic type, if (convergent power series) data are known on a segment of a curve, a power series solution is always known in some neighborhood of each point of the segment.

From (3.2.2) one may show with ease that if f is analytic at a point z_0 (is differentiable in a neighborhood of z_0), then it has all orders of derivatives there. Actually, it can be shown then that the Taylor (power) series

$$f(z) = \sum_{i=0}^{\infty} \frac{f^{(i)}(z_0)}{i!} (z - z_0)^i \qquad (3.2.3)$$

always converges in some neighborhood of z_0. This should strike those who have not studied complex function theory but who have developed an instinct for real functions as too much, because for (3.2.3) to be valid for a function of a real variable requires separately stated conditions that each of the successive derivatives of f exist ($f \in C^\infty$) and still one more condition, that its Taylor series converges! These are, in fact, extremely strong requirements. That to get such features for a function of a complex variable only requires the existence of one derivative in a neighborhood of the point z_0 testifies to the strength of the condition that such a function have a derivative. Actually, as we have seen, it even requires that the real and imaginary parts (of such a function) satisfy the Laplace equation, a rather special and very strong property.

One more comment on the relation to complex variables will be instructive. We define that a real function of a complex variable shall be one where real functional values correspond to real points in the domain. Such functions, if they are analytic, have real Taylor coefficients. The Schwarz reflection principle states that if a real function is analytic in a region D which includes a segment of the real line and if $z = x + iy$ and $\bar{z} = x - iy$ (see Fig. 15) are in D, then

$$\overline{f(\bar{z})} = f(z). \tag{3.2.4}$$

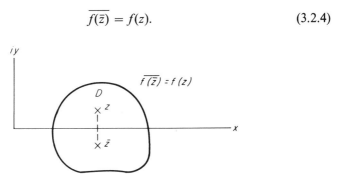

FIG. 15   The Schwarz reflection principle.

This useful law, of course, has a counterpart in potential theory which says that if a function is a solution of the Laplace equation in a region $R \subset E^2$ and is zero on a straight-line segment, then the values of the function agree except for sign at points which are reflected images of each other through this segment. In turn this can be extended to circular arcs (where inverse points, Fig. 16, play the role of reflected points) and to considerations in higher dimensions.

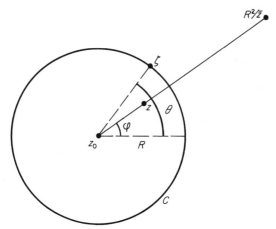

FIG. 16 Poisson's integral comes from the Cauchy integral formula on a circle by separating real and imaginary parts.

3 SOLUTION OF THE DIRICHLET PROBLEM ON A CIRCLE

Assuming a knowledge of (3.2.1) and (3.2.2) we now generate the solution of the Dirichlet problem on a circle $|\zeta| = R$. We use the well-known Cauchy integral formula which already provides such a solution, but the trick is to get the real (or imaginary) part of the solutions in terms of the real (or imaginary) part of the boundary function only.

Let $f(z)$ be analytic on $|z| \leqslant R$ and let $f(\zeta)$ be given. Then for $|z| < R$, we have from (3.2.2) that

$$f(z) = \frac{1}{2\pi i} \oint_C \frac{f(\zeta)}{\zeta - z} \, d\zeta.$$

But then from (3.2.1)

$$\oint_C \frac{f(\zeta)}{\zeta - R^2/\bar{z}} \, d\zeta = 0$$

because $|z| < R$ implies $|R^2/\bar{z}| = R^2/|z| > R$ (so that if z is inside, $R^2/\bar{z}$ is outside the circle; see Fig. 16). It may be noted that $R^2/\bar{z}$ is called the inverse point of z and is regarded as its image or reflected point. Subtracting,

$$f(z) = \frac{1}{2\pi i} \oint_C f(\zeta) \left[\frac{1}{\zeta - z} - \frac{1}{\zeta - R^2/\bar{z}} \right] d\zeta$$

$$= \frac{1}{2\pi i} \oint_C f(\zeta) \left[\frac{z - R^2/\bar{z}}{(\zeta - z)(\zeta - R^2/\bar{z})} \right] d\zeta.$$

We now express ζ and z in polar form,

$$\zeta = Re^{i\theta}, \qquad z = re^{i\varphi}$$

and place them in the formula for $f(z)$ as follows,

$$f(z) = \frac{1}{2\pi} \int_0^{2\pi} f(Re^{i\theta}) \frac{r^2 - R^2}{(Re^{i\theta} - re^{i\varphi})(Re^{i\theta} re^{-i\varphi} - R^2)} Re^{i\theta}\, d\theta$$

$$= \frac{1}{2\pi} \int_0^{2\pi} f(Re^{i\theta}) \frac{R^2 - r^2}{(Re^{i\theta} - re^{i\varphi})(Re^{-i\theta} - re^{-i\varphi})}\, d\theta.$$

Using the De Moivre expression for $e^{i\varphi}$, etc., we obtain an integral representation for the solution of the Dirichlet problem on a circle of radius R,

$$f(re^{i\varphi}) = \frac{1}{2\pi} \int_0^{2\pi} f(Re^{i\theta}) \frac{R^2 - r^2}{R^2 + r^2 - 2Rr\cos(\varphi - \theta)}\, d\theta. \qquad (3.3.1)$$

This is called the Poisson integral, the part under the integral sign multiplying $f(Re^{i\theta})$ being called the Poisson kernel. Note that this kernel is real (as is the factor $1/2\pi$) so (3.3.1) may be rewritten with the real or imaginary parts of $f(re^{i\varphi})$ and $f(Re^{i\theta})$ replacing them, respectively.

4 UNIQUENESS FOR REGULAR SOLUTIONS OF THE DIRICHLET AND NEUMANN PROBLEM ON A RECTANGLE

Just as at the end of Chapter 2 we considered the question of uniqueness of the Dirichlet problem for the wave equation on a rectangle R, we now consider this same question for the Laplace equation (see Fig. 17). Of course, we will obtain a positive result this time.

We have (see Fig. 17),

$$0 = \int_R u(u_{xx} + u_{yy})\, dR$$

$$= \int_R [(uu_x)_x - u_x^2 + (uu_y)_y - u_y^2]\, dR$$

$$= \int_0^y dy \int_0^l (uu_x)_x\, dx + \int_0^l dx \int_0^y (uu_y)_y\, dy - \int_R (u_x^2 + u_y^2)\, dR.$$

From condition (iii) in Section 1 above defining regular solutions, we see that u_x is bounded on $R \cup \delta R$, and since $u = 0$ for $x = 0$ and $x = l$ it follows that the first integral vanishes. Likewise, the second integral vanishes, and this

51

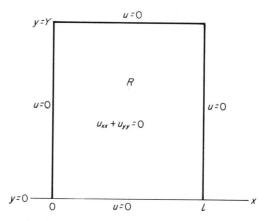

FIG. 17   Uniqueness for the Dirichlet problem for the Laplace equation on a rectangle R.

leaves us with the fact that

$$\int_R (u_x^2 + u_y^2)\, dR = 0.$$

But this gives that in R $u_x = u_y = 0$, and by continuity this is true in $R \cup \delta R$. Then $u = $ constant in $R \cup \delta R$, but on δR, $u = 0$; so it follows that $u = 0$ in $R \cup \delta R$.

The above proof, it will be noted, remains valid right to the last line of argument if the problem we are seeking uniqueness for is one for which the normal derivative of u (u_x on $x = 0, l$; u_y on $y = 0, Y$), rather than u, is specified on R. This is true because the two integrals above were eliminated by virtue of the fact that the product uu_n, not u alone, was zero on the boundary. A problem where the normal derivative is specified on δR is called a Neumann problem. The mixed problem for the Laplace equation is one where u is specified on one portion of the boundary and u_n is specified on the rest of it. Again the above proof is valid right up to the last line. The last line, however, cannot be completed for the Neumann problem because the constant involved cannot be evaluated as zero by reference to a boundary point. The Neumann problem is thus said to be unique up to a constant. For existence of solutions of the Neumann problem, it will turn out that we must require that the integral of the given normal derivative data over the boundary must vanish. This is in the sense of a compatibility relation.

The proof of uniqueness above may be extended quite readily to one in any dimension for any region where the divergence theorem is valid (a Green region), but we will leave the discussion of this point until later.

5 APPROXIMATION METHODS FOR THE DIRICHLET PROBLEM IN E^2

The $\operatorname{Re} z^k$ and $\operatorname{Im} z^k$ are seen to be polynomials in x and y, and since z^k is analytic at any finite point of the real plane, these polynomials are harmonic (are solutions of the Laplace equation) at any finite points (x, y) of the real plane. Let $\{h^k(x, y)\}$ be any family of harmonic polynomials. We determine an approximation u^n to the solution u of a Dirichlet problem on a region A with boundary δA and data f in the following manner.

Let

$$u^n = \sum_{k=1}^{n} c_k h^k(x, y) \tag{3.5.1}$$

be considered an approximation to u if real numbers $c_1, \ldots, c_n$ are chosen so that

$$Q(c_1, \ldots, c_n) = \int_{\delta A} (u^n - f)^2 \, ds \tag{3.5.2}$$

is minimized (where s is arclength). It will be noted that u^n as given in (3.5.1) already satisfies the Laplace equation, and (3.5.2) represents an attempt to fit the boundary data with functions $\{h^k(x, y)\}$ in a sense of least squares. The expression Q in (3.5.2) is a quadratic function of $C_1, \ldots, C_n$, and, moreover, one that is always positive or zero. Thus Q will have just one stationary point, and that will represent a minimum; i.e., we may minimize Q by finding the solution of the set of equations (referred to as normal equations),

$$\partial Q(C_1, \ldots, C_n)/\partial C_h = 0, \qquad h = 1, \ldots, n. \tag{3.5.3}$$

These equations may be written

$$\sum_{k=1}^{n} a_{hk} C_k = b_n \tag{3.5.4}$$

where

$$a_{hk} = \int_A h^k h^h \, dA \qquad \text{and} \qquad b_h = \int_{\delta A} f h^h \, ds. \tag{3.5.5}$$

To obtain our approximation u^n as expressed in (3.5.1) we need only solve (3.5.4) for c_k, $k = 1, \ldots, n$. If all functions h^k are linearly independent, it will be found that the coefficient matrix (a_{hk}) is nonsingular so (3.5.4) does have a unique solution.

We note here in passing that solutions u of the Laplace equation in A have the following property[†] in any number of dimensions,

$$\max_{A \cup \delta A} |u| = \max_{\delta A} |u|. \tag{3.5.6}$$

This extremely important property, known as the maximum (and minimum) principle, will be developed in a later chapter. In the case being treated here of two dimensions, it is, of course, a simple reflection of a well-known theorem for analytic functions of a complex variable. Actually, it is an "a priori inequality" which means that it can be proven without reference to existence considerations and may even be used in the process of proving existence. We will eventually find that such inequalities play a central role in modern analysis.

For now we simply wish to demonstrate that (3.5.6) can be used to determine an error bound in the above least-squares computation. Since in the above both u and u^n are harmonic, then so is $u - u^n$. Then (3.5.6) applied to $u - u^n$ gives

$$\max_{A \cup \delta A} |u - u^n| = \max_{\delta A} |f - u^n|. \tag{3.5.7}$$

Since $c_1, \ldots, c_n$ and f are known, it is at least theoretically possible to compute the maximum appearing on the right-hand side, thus allowing us to judge after the fact if our choice of the family $\{h^k\}$ is adequate to approximate u in this manner according to whatever level of accuracy we have chosen.

Actually, there would often be considerable difficulty in computing the maximum on the right of (3.5.7), especially if high-order harmonic polynomials h^k were to be used. Also, it may not be entirely trivial to solve (3.5.1) to high accuracy if n is large. Altogether one would usually do better to obtain an error bound in $\mathcal{L}^2$ norm, rather than maximum norm, since a minimization is sought in $\mathcal{L}^2$ norm, and also it would be best to be prepared to solve (3.5.4) by iteration to obtain high accuracy. It turns out that the matrix (a_{hk}) is positive definite, a property we find is important and one that derives from early work on least squares. This property makes it possible to use a particularly efficient iterative method in solving (3.5.4). The least-squares method is easily extendable to higher dimensions and to cases where there is no readily available family of approximating functions that solve the differential equation.

Another method of computing an approximation to u consists in laying a rectangular grid over A and writing finite-difference analogs over this grid, regarding the values of u at interior grid points as unknowns while regarding

---

[†] This is not so surprising if specialized to one dimension, for then solutions of the Laplace equation are linear functions and such functions take their extrema at endpoints of any given interval.

values on the boundary as known. This yields a linear algebraic system containing as many unknowns as grid points, and, in order to obtain usable accuracy, the number of unknowns may be in the thousands. Not wanting to write all these equations down, a representative one (involving an index) is written and an iteration is prescribed for solving the system. If the simplest of finite-difference analogs—the five-point analog—is written, one iteration prescribed is successively to form the average at each point of the values at the surrounding points. It happens that this (or an accelerated version of it) is exactly the iteration also preferred for solving the above set of normal equations. Here again the coefficient matrix is positive definite and this iterative method converges. Unfortunately, truly computable error bounds are difficult to come by for finite-difference methods.

The iteration proposed, in which a central value of a five-point cluster is successively replaced by the average of the surrounding values (in some order), gives the $k+1$ estimate, x^{k+1}, of the solution vector x as a linear function of its kth estimate x^k. Thus there is a square matrix M and a column matrix h (the latter representing boundary values) of real numbers so that this iteration may be expressed in the form

$$x^{k+1} = Mx^k + h, \qquad (3.5.8)$$

where x^0 is arbitrary. But, of course, this is the expression (2.5.3) for linear marching schemes, and precisely the same analysis discussed there can be utilized here with one important change. In (2.5.3) we were interested only that the effects of errors in the initial vector not be continually magnified (after some large k) while here we ask that they tend to zero as $k \to \infty$— i.e., that $x^{k+1} \to x$. This requires that we replace our weak inequality condition (2.5.8) with the strong inequality,

$$s = \max_i |\lambda_i| < 1, \qquad (3.5.9)$$

the latter being called the "spectral norm," or more precisely "spectral radius," condition. It is the general (necessary and sufficient) criterion for convergence of all linear iterative techniques (3.5.8).

Let the linear system arising either from least squares [see (3.5.4)] or from five-point finite-difference methods of approximation be written,

$$Ax = (L+D+U)x = b \quad \text{or} \quad \sum_{j=1}^{n} a_{ij} x_j = b_i \qquad (3.5.10)$$

where L is a lower diagonal, D a diagonal, and U an upper diagonal matrix. Suppose that for all i, $a_{ii} \neq 0$. The method spoken of above, whether applied to least-squares or finite-difference methods of approximation, corrects the ith component of the solution vector x from the ith equation, proceeding from one equation to the next in successive steps and then repeating. For purposes of analysis only, this computation may be written

$$x^{k+1} = -(L+D)^{-1}Ux^k - (L+D)^{-1}b. \qquad (3.5.11)$$

A basic theorem states that if A is positive definite $(a_{ij} = a_{ji}$, and $x \neq 0$ implies $x^T A x > 0$), then $M = -(L+D)^{-1}U$ has spectral norm less than one. Moreover, under the same conditions, a very successful accelerated form has a correction matrix $M(\omega)$ (here ω is the acceleration parameter) which has spectral norm less than one for $0 < \omega < 2$. The method became popular under the name overrelaxation [40], starting from thesis work of Young, [46], but we have tended to prefer a more descriptive term, accelerated successive replacements [3, 28]. As stated, both the coefficient matrix of least squares [see (3.5.4)] and the coefficient matrix for the five-point finite-difference analog are positive definite.

EXERCISE 1   From (3.5.5) show that (a_{hk}) is positive definite.

EXERCISE 2   Show that the linear system in the five-point finite-difference analog of the Laplace equation has a positive-definite coefficient matrix.

6 THE CAUCHY PROBLEM FOR THE LAPLACE EQUATION

In the spirit adopted in this outline, of exhausting all interesting possibilities of boundary-value problems, at least in E^2, we now turn our attention to the Cauchy problem for the Laplace equation. If one expects that the world exhibits a kind symmetry with respect to boundary-value problems, he will be surprised to find that this problem is not as bizarre as the Dirichlet problem for the wave equation.

The problem is

$$\psi_{xx} + \psi_{yy} = 0$$

$$\psi(x,0) = f(x), \qquad \psi_y(x,0) = g(x)$$

where f and g are defined on $(a, b) \subset R^1$. Obviously, all regular solutions are given by replacing y by iy in solution of the similar problem for the wave equation given in Chapter 1. That is, the D'Alembert solution here becomes

$$\psi(x,y) = \tfrac{1}{2}f(x+iy) + f(x-iy) - \tfrac{1}{2}i \int_{x-iy}^{x+iy} g(\xi)\,d\xi. \qquad (3.6.1)$$

It has thus become necessary to extend the domain of f and g to some region R of the complex plane containing (a, b). In order that $\psi \in C^2$ in some region of E^2 containing (a, b) we ask that $f \in C^2(R)$ and $g \in C^1(R)$. But since R is in the complex plane, our functions f and g must be analytic there. Then, since a quick argument shows that our solution (3.6.1) is unique, we find that regular solutions for ψ are analytic, and moreover, that a solution of our initial value problem can exist in E^2 in neighborhoods of points of (a, b) only if f and g are analytic in (a, b)!

This is a much stronger requirement to make of given functions f and g than we had to require for the data on the wave equation where we only required that $f \in C^2$ and $g \in C^1$. The remarkable importance of this difference becomes apparent when we consider problems of stability, for it is evident that here only analytic perturbations of f and g can be tolerated while a much wider class of perturbations can be tolerated for the corresponding wave equation problem. These matters will be discussed again later in this outline (Chapter 5, Section 2). It has been recognized only in very recent times that the difficulty is not so devastating as to make the Cauchy problem for elliptic type useless.

Using the Schwarz reflection principle (see Section 2) we now rewrite (3.6.1) in a more compact form. Taking a path $x = $ constant for the complex integral,

$$\psi(x, y) = \operatorname{Re} f(x+iy) - \tfrac{1}{2}i\left[\int_x^{x+iy} g(\xi)\, d\xi - \int_x^{x-iy} g(\xi)\, d\xi\right]$$

$$= \operatorname{Re} f(x+iy) - \tfrac{1}{2}i\left[i\int_0^y g(x+it)\, dt + i\int_0^y g(x-it)\, dt\right]$$

$$= \operatorname{Re} f(x+iy) + \tfrac{1}{2}\int_0^y [g(x+it) + g(x-it)]\, dt$$

or

$$\psi(x, y) = \operatorname{Re} f(x+iy) + \int_0^y \operatorname{Re} g(x+it)\, dt. \qquad (3.6.2)$$

It should be noted that the solution (3.6.2) is not necessarily valid in the entire (x, y) plane. Rather, if f and g have convergent Taylor series on the data segment, it can be said that the solution (3.6.2) is valid in some (x, y) neighborhoods of points of the data segment. Just where it is valid will be discussed later. In order for it to be valid in the entire plane (or half plane), data would have to be specified on the entire x axis, not just on a segment. Moreover, a relatively simple transformation takes our solution in Section 3 for the Dirichlet problem with data u (only) on a circle into the solution of the

Dirichlet problem for the upper half-plane with data u (only) on the x axis; i.e., there is a simple harmonic-preserving (conformal) mapping that maps the boundary of the circle into one of its tangent lines and maps the interior of the circle into a half plane above this line. It is thus seen that the solution of the Cauchy problem, if it is valid in the entire upper half-plane, will necessarily correspond to data $u_y = g$ which is superfluous since $u = f$ alone is enough to determine the solution. To find a function $u_y = g$ which would work, one could solve the Dirichlet problem corresponding to prescription of u alone and then simply compute u_y on the line $y = 0$. Thus a very strong compatibility relation will have to be satisfied on data f and g in order that (3.6.2) be valid in an entire half-plane.

A similar situation should be described with respect to the Dirichlet problem for any bounded region. One could always propose an exterior problem. For example, in Fig. 13 it could be required to solve for u in the region exterior to δA, where the Laplace equation is then assumed to be satisfied, only, of course, some prescription on the behavior "in a neighborhood of infinity" must be given. To see the nature of this prescription, the problem could always be transformed into an interior problem by performing an inversion at a circle (see Fig. 18) which lies wholly within the interior region A. For higher n-dimensional problems, a similar process can be followed by performing a Kelvin transformation (inversion at a sphere followed by multiplication with $1/r^{n-2}$). This will be mentioned again in Chapter 10 on potential theory.

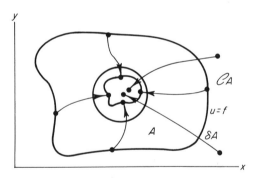

FIG. 18   Inversion at a circle is harmonic preserving.

4

Various Boundary-Value Problems
for Simple Equations of Parabolic Type

1 THE SLAB PROBLEM

A function u will be said to be a regular solution of the slab problem for the heat equation on a rectangle (see Fig. 19)

$$R = \{(x, t) | 0 < x < l, 0 < t < T\}$$

if

(i) $u \in C^2(R)^\dagger$;
(ii) $u_{xx} - u_t = 0$ in R;
(iii) $u \in C^1(R \cup \delta R)^\dagger$; and
(iv) $u(x, 0) = f(x)$, $u(0, t) = g_0(t)$, $u(l, t) = g_l(t)$.

This problem is much like the vibrating string problem except that primitive data only, not derivative data, are given on the initial line $t = 0$. This is to be expected because the differential equation contains one less derivative with respect to t. Also, it is to be expected that the line $t = T$ is a naturally occurring boundary since it is a characteristic. Just why data should be given on

---

† These may be weakened to: u_{xx} and u_t are continuous in R, u_x and u are continuous in $R \cup \delta R$. We have here preserved a similarity to the vibrating string problem, but this puts an unnecessarily strong restraint.

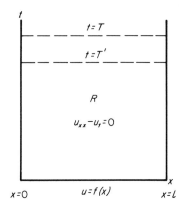

FIG. 19   The slab problem for the heat equation.

$t = 0$, and not on $t = T$, instead of vice versa is not yet clear, but this point will be taken up.

First, let us verify our expectations by proving uniqueness for regular solutions of slab problems as posed. Taking a variation of the proof used to establish uniqueness of the Dirichlet problem for the Laplace equation and uniqueness of the vibrating string problem, we put

$$R' = \{(x, t) | 0 < x < l, \quad 0 < t < T'\} \qquad \text{where} \quad 0 < T' \leqslant T.$$

Then,

$$0 = \int_{R'} u(u_{xx} - u_t)\, dR = \int_{R'} [(uu_x)_x - u_x^2 - \tfrac{1}{2}(u^2)_t]\, dR$$

$$= \int_0^{T'} dt \int_0^l (uu_x)_x\, dx - \tfrac{1}{2} \int_0^l dx \int_0^{T'} (u^2)_t\, dt - \int_{R'} u_x^2\, dR$$

$$= 0 - \tfrac{1}{2} \int_0^l u^2(x, T')\, dx - \int_{R'} u_x^2\, dR.$$

It follows that for every $0 < T' \leqslant T$, $u(x, T') = 0$ for every $0 < x < l$; i.e., it follows that $u = 0$ on R. Of course, here we have used that $u(0, t) = u(l, t) = u(x, 0) = 0$ as required by the uniqueness problem. Moreover, the proof obviously remains valid if the normal derivatives are specified on the side boundaries $x = 0, l$.

Physically here we think of an infinite slab where x is a thickness parameter, t is time, and the faces are heated uniformly. The initial temperature distribution through the slab $u(x, 0)$ is known and either the temperatures of the faces are kept at known values $u(0, t)$, $u(l, t)$ or the heat flux $u_x(0, t)$, $u_x(l, t)$ on the faces is prescribed. This latter would often be the insulated case, $u_x(0, t) = u_x(l, t) = 0$.

Perhaps an easier physical realization to picture for the problem is that of a thin rod with variable length x where the initial temperature distribution on the rod is known, and either the temperature is controlled on the ends or the heat flux at these ends is prescribed. Of course, in both cases it is essentially required to find the temperature at all points at all future times, but one may sometimes be satisfied with less. For example, one may only need to know if the temperature at any point on the rod will exceed the melting point at any time. It is possible to answer this question without really solving the problem because there is a maximum principle which says that u takes on its maximum at a point on that part of the boundary where the data is specified.

2 AN ALTERNATIVE PROOF OF UNIQUENESS

A modification of the uniqueness proof in Section 1 is the following. Let

$$I(t) = \tfrac{1}{2} \int_0^l u^2(x, t)\, dx. \tag{4.2.1}$$

Then

$$I'(t) = \int_0^l u u_t\, dx = \int_0^l u u_{xx}\, dx = \int_0^l (u u_x)_x - u_x{}^2\, dx = -\int_0^l u_x{}^2\, dx.$$

From $u(x, 0) = 0$ we have that

$$I(0) = 0. \tag{4.2.2}$$

However, the above shows that for every $t \geqslant 0$,

$$I'(t) \leqslant 0. \tag{4.2.3}$$

This implies that for every $t \geqslant 0$,

$$I(t) \leqslant 0. \tag{4.2.4}$$

But from (4.2.1), we have that $I(t) \geqslant 0$, so that it is seen that $I(t) = 0$, from which, of course, $u = 0$ on R.

The nature of this proof will be important to us, since (4.2.3) gives some interesting results about the asymptotic behavior of the difference of two solutions even if the two solutions do not correspond to the same initial values as required by (4.2.2).

3 SOLUTION BY SEPARATION OF VARIABLES

As indicated, our curiosity should now be aroused about why data are specified on $t = 0$, and not on $t = T$, rather than vice versa.[†]

In this connection it will be instructive first to give a formal development of the solution of the problem with $g_0(t) = g_l(t) = 0$, $f(x) = \sum a_n \sin nx$ by separation of variables. From $u = \varphi(x)\psi(t)$ we have from the heat equation

$$\varphi''(x)/\varphi(x) = \psi'(t)/\psi(t),$$

which we put equal to $-k^2$. The negative constant is chosen so that φ may satisfy the side data. We have

$$\varphi(x) = A \sin kx + B \cos kx$$

or

$$0 = A \cdot 0 + B \Rightarrow B = 0$$

and

$$0 = A \sin kl \Rightarrow k = \pi n/l, \qquad n = 1, 2, \dots.$$

Then

$$\psi(t) = C \exp(-\pi^2 n^2 t/l^2),$$

so, in total

$$u(x, t) = \sum_{n=1}^{\infty} A_n \exp(-\pi^2 n^2 t/l^2) \sin(\pi n x/l), \qquad (4.3.1)$$

where it is now evident that to satisfy the initial data we should select the A_n to be the given coefficients a_n in the Fourier series for f. In other treatises it is shown that (4.3.1) converges under very general conditions of f.

4 INSTABILITY FOR NEGATIVE TIMES

It will be noted now that small errors or changes in the initial data f are quickly damped out in time according to (4.3.1) because although the coefficients A_n or a_n may change, this change is multiplied by a decaying exponential. Now consider the problem of finding solutions for $t < 0$ [with

---

[†] Obviously to specify data on both lines would be too much—and we could not expect a solution to exist unless a strong compatibility relation was stated—because we have proved uniqueness with data specified on $t = 0$ only.

the same boundary conditions $u(0, t) = u(l, t) = 0$]. This is accomplished by replacing t in (4.3.1) by $-t$, but it will then be seen that small errors or changes in initial data are exponentially amplified. The problem is said to be asymptotically unstable. It should now be evident why u is not specified on $t = T$ instead of on $t = 0$.

If the heat conductivity of the rod or slab material is K, then the equation of physical significance is $(Ku_x)_x = u_t$. When K is constant, it can be absorbed as a scale in x, which is the way we have here treated the problem.

5 CAUCHY PROBLEM ON THE INFINITE LINE

If initial data alone are to be given on some line $t = 0$ for the heat equation, we can expect that only one piece of data will be given because $t = 0$ is a characteristic for the heat equation. Moreover, in view of the fact that solutions do exist for the slab problem it is evident that one cannot simply give u on a segment of the line $t = 0$ and expect to be able to establish uniqueness. It is thus reasonable that, for a Cauchy problem on the line $t = 0$, one should expect to specify u on the infinite line.

Infinite rods are hard to come by, but tubes containing fluid in a virtually stagnant condition with lengths sufficiently long with respect to diameter that they may be treated as infinitely long, abound in nature and technology. It happens that the equation

$$DU_{xx} = u_t \tag{4.5.1}$$

again applies where u is the concentration of a given substance (say salt) in the fluid (say water) and D is the coefficient of diffusion. As an ideal diffusion problem we think of placing a finite mass C_0 of salt at a point $x = 0$ in a straight channel, thus giving rise to an infinite concentration u at the point $x = 0$. Thus we pose the following problem for (4.5.1):

$$\text{at} \quad x = 0, \qquad \lim_{t \to 0} u = \infty, \tag{4.5.2}$$

$$\text{at} \quad x \neq 0, \qquad \lim_{t \to 0} u = 0, \tag{4.5.3}$$

and

$$\text{for every} \quad t > 0, \qquad \int_{-\infty}^{\infty} u \, dx = C_0 \in R^1, \tag{4.5.4}$$

the latter being a mass-conservation condition.

It can readily be shown that (4.5.1)–(4.5.4) determine a unique limit of a sequence of unique regular solutions. A regular solution here is one that has two continuous derivatives in x and one continuous derivative in t in the (open) upper half-plane and is continuous in the upper half-plane joined with the line $t = 0$. We also assume that u_x is bounded in the upper half-plane. Consider a sequence of problems with initial functions defined by (4.5.2) and (4.5.3) replaced by a sharpening, heightening sequence of bell-shaped functions; the meaning of the problem in the first place can be regarded as requiring that we look at the limit of solutions of such a sequence of problems. For any one of them, suppose $u = u_1 - u_2$ is the difference of any two regular solutions; we show that u must be zero. Note that from (4.5.4) $\lim_{x \to \pm \infty} u = 0$. Taking a clue from our alternative proof of uniqueness (Section 2) of the slab problem, we let

$$I(t) = (1/2D) \int_{-\infty}^{\infty} u^2 \, dx. \tag{4.5.5}$$

Then

$$I'(t) = (1/D) \int_{-\infty}^{\infty} uu_t \, dx = \int_{-\infty}^{\infty} uu_{xx} \, dx$$

$$= \int_{-\infty}^{\infty} ((uu_x)_x - u_x^2) \, dx = -\int_{-\infty}^{\infty} u_x^2 \, dx \leqslant 0.$$

As before, since $I'(t) \leqslant 0$ and $I(0) = 0$, it must follow that $I(t) \leqslant 0$, but (4.5.5) has it that $I(t) \geqslant 0$ from which we conclude that $I(t) = 0$ and, since u is regular, that $u = 0$ for every x and for $t \geqslant 0$.

The solution to the diffusion problem (4.5.1)–(4.5.4) is

$$u(x, t) = [C_0/2(\pi Dt)^{1/2}] \exp(-x^2/4Dt), \tag{4.5.6}$$

a fact apparently known to Fourier.

EXERCISE 1 Verify that (4.5.6) satisfies (4.5.1).

EXERCISE 2 Verify that (4.5.6) satisfies (4.5.2) and (4.5.3).

EXERCISE 3 Verify that (4.5.6) satisfies (4.5.4).

Hint Use the well-known infinite integral of the error function.

In our study of transmission lines later it will be found that the same problem elucidates the behavior of pure resistance–capacitance lines.

Of course, as one might clearly expect, a mixture of the Cauchy and slab problems is possible where data are given on the half infinite x axis and on a finite or infinite segment of the t axis, the lower end of which touches the x axis.

EXERCISE 4 Define regular solutions for this mixed problem and prove uniqueness.

6 UNIQUE CONTINUATION

Now we ask about prescription of a Cauchy problem on a *non*characteristic line, say on $x = 0$. Here we would expect to prescribe two (not one) functions, u and u_x, on a finite segment (not on an infinite line as in Section 5), and since there is just one family of characteristics (which do not intersect or coalesce) we would expect the solution to be determined in an infinite strip. After all, the characteristics do still form "natural barriers."

Let $u_{xx} = u_t$ and

$$u(0, t) = f(t), \qquad u_x(0, t) = g(t) \tag{4.6.1}$$

for $f, g: (a, b) \to R^1$ (see Fig. 20). The solution is

$$u(x, t) = \sum_{i=0}^{\infty} \frac{x^{2i}}{(2i)!} f^{(i)}(t) + \frac{x^{2i+1}}{(2i+1)!} g^{(i)}(t) \tag{4.6.2}$$

as it appears in footnote 1 on page 446 of the exhaustive classical work of Gevrey [17] on parabolic type.

EXERCISE 5 Determine conditions on f and g so that the two series appearing in (4.6.2) converge.

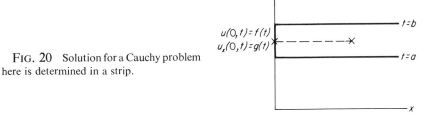

FIG. 20 Solution for a Cauchy problem here is determined in a strip.

7 POISEUILLE FLOW

The problem of flow of an incompressible fluid through a tube with rigid walls in response to a pressure gradient down the length of the tube will now be considered (see Fig. 21). Our differential equations arise from equating the

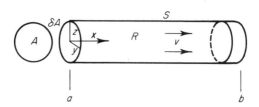

FIG. 21   A rigid-wall tube.

inertial $F = ma$ forces in the x direction minus the drag force F_D in the x direction, due to viscous shear, to the pressure force $F_p = -p_x$ and dividing by the mass per unit volume. The pressure force is a surface force, and we require a difference or derivative of pressure to create a body (or volume) force capable of moving the fluid—capable, that is, of being balanced against the inertial force which is a body force. Where d is density, we write

$$(1/d) F_D - (1/d) F = -(1/d) F_p$$

and let the pressure gradient be written

$$p_x = -F(x, t).$$

Then velocity can be written $v = v(x, y, z, t)$ where t is time. Acceleration is

$$Dv/Dt = (d/dt) v(x(t), y(t), z(t), t)$$

$$= v_t + (dx/dt) v_x + (dy/dt) v_y + (dz/dt) v_z \qquad (4.7.1)$$

where $x = x(t)$, $y = y(t)$, $z = z(t)$ describe a time trajectory of an element of fluid. On a trajectory for which the velocity is always pointed in the x direction, we have $dy/dt = dz/dt = 0$ and $dx/dt = v$, so the acceleration is given by

$$Dv/Dt = v_t + vv_x. \qquad (4.7.2)$$

The equation of motion in the x direction is then found to be

$$(\mu/d)(v_{xx} + v_{yy} + v_{zz}) - v_t - vv_x = -F(x, t)/d. \qquad (4.7.3)$$

66

Here the shear stress between layers of fluid has been assumed to be proportional to the differential of velocity normal to their direction of motion, where μ is the proportionality factor called the viscosity coefficient. If layers traveled at the same rate, there would be no shear between them, but not all layers can travel at the same rate because the walls are fixed. The second derivatives of velocity appear in the equation of motion because shear stress is a surface force, and, like pressure, requires a differential change across an element in order to supply a net force which can be used in the equation of motion where forces on a volume (or points) are equated.

Flows of incompressible fluids in rigid tubes exhibit no local pressure gradients that are different than that given by the pressure drop (per unit length) across the length of the entire tube. If the pressure is raised at some point in the tube, there is in this idealization no relief available in the axial direction because the fluid is assumed incompressible ($\partial d/\partial p = 0$), and there is no relief laterally either because the walls are not compliant. The only thing that can happen then is that on raising the pressure at one point an increased pressure gradient is installed instantaneously down the whole tube and across it. Theoretically, changes of pressure transmit instantaneously down a rigid channel filled with incompressible fluid. No compression and rarefaction wave can be thought of as traveling down such a channel; the density is assumed constant. One thinks simply of two large reservoirs filled with the same liquid maintained as different pressures even though this liquid flows down a connecting tube between them. If the pressure is changed in one reservoir (only), the pressure gradient must be assumed changed down the tube instantaneously. To acknowledge that realistically this would not be so would be to include effects of compressibility and wall compliance. Of course, since there are no local pressure gradients different than the overall gradient, and pressure is the only driving force in the fluid, it follows that the velocity, in say a circular cylinder tube, does not vary down the channel but only across channel. Thus we put

$$v = v(y, z, t); \qquad -p_x = F(x, t) = f(t).$$

Then in Eq. (4.7.3), $v_x = v_{xx} = 0$ so the equation of motion becomes

$$(\mu/d)(v_{yy} + v_{zz}) - v_t = -f(t)/d. \qquad (4.7.4)$$

For channels of radius r with small Reynold's number dvr/μ, the effects of viscosity must be considered important, and the fluid is assumed "to stick" to the walls. If S is the inner channel lateral surface, we have

$$v = 0 \quad \text{on} \quad S. \qquad (4.7.5)$$

67

A regular solution of tube flow will be a function v defined on A such that

(i) for every t, v has two continuous derivatives with respect to y and z in A;

(ii) for every t, v has one continuous derivative with respect to y and z in $A \cup \delta A$;

(iii) for every t, v has one continuous derivative with respect to t at every point in A; and

(iv) v satisfies the differential equation (4.7.4) and boundary condition (4.7.5).

Assuming a tube which is a circular cylinder, then, we may put $\rho = (y^2 + z^2)^{1/2}$ and (4.7.4) becomes

$$v_{\rho\rho} + (1/\rho)v_\rho - (1/v)v_t = -f(t)/\mu \qquad (4.7.6)$$

where $v = \mu/d$. From this we have

$$(1/\rho)(\rho v_\rho)_\rho = -f(t)/\mu + (1/v)v_t. \qquad (4.7.6a)$$

Integrating from 0 to ρ and noting from condition (ii) above that v_ρ is bounded in A,

$$\rho v_\rho = -(f(t)/2\mu)\rho^2 + (1/v)\int_0^\rho \rho_1 v_t(\rho_1, t)\, d\rho_1.$$

Dividing by ρ and integrating from ρ to r, and using again the boundary condition (4.7.5), we have

$$v(\rho, t) = (f(t)/4\mu)(r^2 - \rho^2) - (1/v)\int_\rho^r (1/\rho_2)\, d\rho_2 \int_0^{\rho_2} \rho_1 v_t(\rho_1, t)\, d\rho_1. \quad (4.7.7)$$

This equation is designed so that v can be obtained by iteration (in much the way that we will eventually prove the existence and uniqueness of solutions of boundary-value problems for the nonhomogeneous wave equation), starting with the term free of integration. This is successfully carried out in our monograph [29] for the case where f is analytic. For the purposes of this outline, we consider only the classical case where f is constant. Then iterating (4.7.7) with the free term as the initial iterate we immediately obtain the term

$$v(\rho, t) = -(p_x/4\mu)/(r^2 - \rho^2) \qquad (4.7.8)$$

as our solution where $-p_x$ is the difference of outlet to inlet pressure divided by the length of the tube. Integrating over A we obtain

$$Q = 2\pi \int_0^r \rho v(\rho, t)\, d\rho = 2\pi(-p_x)r^4/8u, \qquad (4.7.9)$$

which is the fourth-power law for volume flow rate observed by Poiseuille according to his manuscript of 1836 [35].

8 MEAN-SQUARE ASYMPTOTIC UNIQUENESS

If to (4.7.5) and (4.7.6) we were to add a prescription of the initial velocity, $v(\rho, 0) = g(\rho)$, we would have a slab-type problem closely related to the problem of Section 1, 2, and 3. Evidently the same techniques as used in Sections 1 and 2 could be utilized to prove uniqueness of this problem.

EXERCISE 6 Prove uniqueness of the initial boundary-value problem (4.7.5), (4.7.6), and $v(\rho, 0) = g(\rho)$.

Hint Use (4.7.6) in the form (4.7.6a).

But the condition $v(\rho, 0) = g(\rho)$ would not seem to have any relevance to the result (4.7.9),[†] a well-observed fact of life for small Reynolds numbers. Actually, since the viscous shear stress represents a damping, it could certainly be expected that the effect of any initial velocity would be quickly washed out. In fact, this turns out to be the truth.

Let

$$I(t) = (1/2v) \int_A (v-w)^2 \, dA = (1/2v) \int_A u^2 \, dA$$

where $u = v - w$ is the difference of two solutions corresponding possibly to different initial conditions. Using the more general differential equation (4.7.4), putting

$$uu_{xx} + uu_{yy} = (uu_x)_x + (uu_y)_y - u_x^2 - u_y^2,$$

and using the divergence theorem[‡] on the vector field (uu_x, uu_y),

$$I'(t) = \tfrac{1}{2} \int_A uu_t \, dA = \int_A u \, \Delta u \, dA = \int_S uu_n \, ds - \int_A (u_y^2 + u_z^2) \, dA,$$

and, by the boundary condition (4.7.5), the first term can be dropped so that

$$I'(t) = -\int_A (u_y^2 + u_z^2) \, dA. \tag{4.8.1}$$

But the Raleigh–Ritz principle, which we will take up in Chapter 10, states that the first positive eigenvalue associated with the Dirichlet problem in A

---

† Even if $g(\rho) = 0$.
‡ $\int_A \operatorname{div} \mathbf{a} \, dA = \int_A (\mathbf{a}, \mathbf{n}) \, ds$ where $\mathbf{a}$ is a vector field. See Chapter 10 on potential theory. $u_n = (\operatorname{grad} u, \mathbf{n})$.

for the Laplace equation [i.e., the first positive number λ^2 such that the boundary-value problem

$$\varphi_{xx} + \varphi_{yy} + \lambda^2\varphi = 0 \quad \text{in} \quad A, \qquad \varphi = 0 \quad \text{in} \quad \delta A,$$

admits a solution φ such that $\varphi(x,y) \neq 0$ for some (x,y) in A] is given by

$$\lambda^2 = \text{glb}\frac{\int_A(\varphi_x{}^2 + \varphi_y{}^2)\,dA}{\int_A\varphi^2\,dA}, \tag{4.8.2}$$

where the greatest lower bound is taken over functions φ that are zero in S. Again using the boundary condition (4.7.5), we have

$$\lambda^2 \leqslant \frac{\int_A(u_x{}^2 + u_y{}^2)\,dA}{\int_A u^2\,dA}$$

or

$$-\int_A(u_x{}^2 + u_y{}^2)\,dA \leqslant -\lambda^2\int_A u^2\,dA.$$

Then from (4.8.1)

$$I'(t) \leqslant -\lambda^2\int_A u^2\,dA = -2\nu\lambda_r{}^2 I(t). \tag{4.8.3}$$

Here, of course, λ_r is the first eigenvalue associated with the Dirichlet problem for the Laplace equation on the circle A. It can be shown that

$$\lambda_r = \mu_0/r \tag{4.8.4}$$

where μ_0 is the first positive root of the Bessel function $J_0(\mu)$.

EXERCISE 7 Prove (4.8.4).

Then (4.8.3) becomes

$$I'(t) \leqslant -2\mu_0{}^2(\nu/r^2)I(t)$$

or

$$(\exp[2\mu_0{}^2(\nu/r^2)t]\,I(t))' = \exp[2\mu_0{}^2(\nu/r^2)t]\,I'(t)$$
$$+ 2\mu_0{}^2(\nu/r^2)\exp[2\mu_0{}^2(\nu/r^2)t]\,I(t) \leqslant 0,$$

and, therefore, $\exp[2\mu_0{}^2(\nu/r^2)t]\,I(t)$ is bounded by a positive constant or zero. We have then that

$$I(t) \leqslant \exp[-2\mu_0{}^2(\nu/r^2)t]\,I(0). \tag{4.8.5}$$

This shows that the mean-square difference of any two regular tube flows is bounded by a decaying exponential, often one with a large exponential rate (if μ is large and r is small). It is this effect that we call mean-square asymptotic uniqueness. We may write (4.8.5) as

$$I(t) \leqslant \exp[-2\mu_0^{\,2}(\nu/r^2\mathscr{F})\,t/T]\,I(0) = \exp[-2(\mu_0/\alpha)^2\,t/T]$$

where T is a period of pressure gradient, $\mathscr{F} = 1/T$, and α^2 is the dimensionless parameter $(r^2/\nu)\mathscr{F} = (r^2 d/\mu)\mathscr{F}$.

EXERCISE 8 Repeat the above proof for the special case of axially symmetric flow [i.e., using (4.7.6) rather than (4.7.4)].

9 SOLUTION OF A DIRICHLET PROBLEM FOR AN EQUATION OF PARABOLIC TYPE

The reader will now be curious as to what extent the Dirichlet problem is bizarre for equations of parabolic type. Obviously, from Sections 1 and 2 the prescription of a Dirichlet problem for the heat equation would require a strong compatibility condition on the prescribed function. We develop here, however, the solution of a Dirichlet problem for another equation of parabolic type.

The problem is posed on the two-unit square S; (see Fig. 22). In S,

$$y^2 u_{xx} - 2xy u_{xy} + x^2 u_{yy} - x u_x - y u_y - u = 0, \qquad (4.9.1)$$

$$u(x, -1) = u(x, +1) = x^2, \qquad u(-1, y) = u(1, y) = y^2 \qquad (4.9.2)$$

Exercise 6, Chapter 1 showed that this equation has the circles concentric to the origin as its single family of characteristics. When we put (4.9.1) in polar

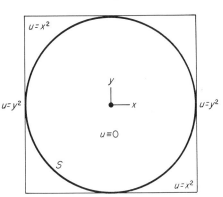

FIG. 22 Dirichlet problem for an equation taken from a theory of Gaetano Fichera.

coordinates (ρ, θ) we obtain the equation

$$u_{\theta\theta} - u = 0, \tag{4.9.3}$$

an equation free of ρ. On any *complete* circular path C in S, we find that solutions of (4.9.3) must be periodic, but all solutions of (4.9.3) being real exponentials, this is possible only if $u \equiv 0$ on C. Thus we have that $u \equiv 0$ inside the largest circle interior to S. For those circular paths in the corners, we must solve simple linear two-point boundary-value problems. We find then that

$$u = \begin{cases} 0 & \text{for} \quad x^2 + y^2 \leqslant 1, \\ (x^2 + y^2 - 1)\left(\dfrac{\exp \arcsin[|x|/(x^2+y^2)^{\frac{1}{2}}] + \exp \arccos[|y|/(x^2+y^2)^{\frac{1}{2}}]}{\exp \arcsin[1/(x^2+y^2)^{\frac{1}{2}}] + \exp \arccos[1/(x^2+y^2)^{\frac{1}{2}}]} \right) \\ \hspace{8cm} \text{for} \quad x^2 + y^2 > 1. \end{cases} \tag{4.9.4}$$

EXERCISE 9 Check the manipulations yielding (4.9.4).

EXERCISE 10 Show that u_ρ does not exist on the curve $\rho = 1$ for the solution quoted in (4.9.4).

Parabolic equations live in a world of their own and deserve quite as much attention as the other two types. We have heard it contended that equations of parabolic type are simply a transition or degenerate case; the two characteristics of hyperbolic type approach a limiting position, folding out or in, forming parabolic type in transition to elliptic type. This is the impression one gets from a study of the equations of inviscid flow where only the sonic line is a "region" of parabolicity; it is also the impression of transmission line theory where the equations become parabolic type if one forgets to include effects of inductance.

However, there are theories of partial differential equations where equations of parabolic type are made to play a basic role. Equations such as $u_t = Lu$, where Lu is a general partial differential operator including elliptic and hyperbolic type, are studied. Here, when u is taken independent of t, this gives theories of elliptic and hyperbolic type. Sometimes this even seems to be physically the most natural way to study such problems. We have shown in this chapter an extensive array of interesting problems for parabolic type, and Part IV of this text will develop a complete theory of elliptic–parabolic type which is clearly in the nature of a generalization of parabolic type.

We may mention that equations of parabolic type govern physical phenomena involving heat conduction, blood flow, kidney mechanisms, electrophoresis, boundary-layer theory, and energy flux in the evolution of stellar interiors.

5

Expectations for Well-Posed Problems

1 SENSE OF HADAMARD

As presented more than 50 years ago by the late great analyst–universalist, Jaques Hadamard, a boundary-value problem is said to be well posed if it has been proved that (i) a solution exists, (ii) the solution is unique, and (iii) the solution is continuously dependent on its data. The modern analyst finds these phrases to be somewhat misleading because their meanings have been greatly expanded, or even altered, by the impact of modern analysis, especially functional analysis. There is even as yet no clear agreement as to the meanings we should accept for these statements, and it seems to become clear now that for different purposes, different meanings might be useful. We will first explore Hadamard's meaning, which is certainly still quite valid for many purposes, and then indicate what changes have developed in modern thinking. Other than the very active field built around methods of approximation to solutions of boundary-value problems which, when it envisages the use of computers, is called numerical analysis, the study of partial differential equations has come to be devoted to the exploration of ideas related to existence and uniqueness.

Until perhaps 15 to 20 years ago the statements above were taken in the following sense:

(i) By existence was meant existence of regular solutions. Since there was no other sense explicitly expounded, this was not always as explicitly stated as it might have been, but this was generally the intent. The properties of a regular solution were explicitly and carefully put down by the most thorough writers, but as we have seen in Chapters 2–4, the precise specification of what it is one intends by regular solutions often takes the statement of slightly different conditions for different problems, and this may have lead to some confusion. It leads to some confusion now in the reading of literature that is even only 20 to 25 years old! Some examples, such as the impulse problem for the heat equation, as shown here in Chapter 4, Section 5, long indicated a need for at least a minor extension in the meaning of existence of solutions and finally lead to it, but only after many generations. In some moods modern analysts have tended to feel that we were simply too enamored with the genius of Lagrange and allowed him to mislead us in this direction, but this would seem to need further reflection. Specifically, the trouble with our concepts of regular solutions is that it has been found to be difficult, or impossible perhaps, to prove existence for large classes of solutions of boundary-value problems using these concepts.

(ii) By uniqueness was meant uniqueness of regular solutions. It was almost always easy, just as we have seen it to be in the very elementary problems solved in Chapters 2, 3, and 4, to prove uniqueness in these senses, and this apparently represents the "survival value" of the concept of regular solutions in the evolution of mathematical thought, at least from a purely mathematical point of view.[†] When the sense of existence is weakened, of course, this automatically weakens the sense of uniqueness demanded, and workers in the field run the risk that when they have found a seemingly reasonable sense of existence which is weak enough that it allows them to prove existence for large classes of boundary-value problems, it may prove too weak in that it allows for a multiplicity of solutions. This has happened, as a tracing of recent literature will reveal. So far as we know, a direct weakening of the sense of uniqueness has never been attempted, except in our sense of mean-square asymptotic stability as expounded in Chapter 4, Section 8. We contend that this technique, when it can be utilized, is often more meaningful for physical problems.

(iii) The concept of continuous dependence was introduced to eliminate from our considerations those boundary-value problems for which the solutions are thought to be not meaningful physically by virtue of an instability with respect to prescribed data. Unfortunately, Hadamard's precise description of this property was too abstract from a physical point of view to be

---

[†] We will see that in physical interpretation of results, the regular property of solutions may well be indispensible in many cases.

readily acceptable by physicists and engineers. Hadamard stated that the solution of a boundary-value problem is continuously dependent on its data if the sequence of unique (regular) solutions corresponding to a uniformly convergent sequence of data converges to a solution of the boundary-value problem corresponding to the limit function of the sequence of data.

In this sense, the Cauchy problem for the homogeneous wave equation (in E^2) is well posed, but that for the Laplace equation is not. We will discuss the point about the Laplace equation here later; for now we verify the statement about the wave equation. Let $\{f_n(x)\} \to f(x)$ and $\{g_n(x)\} \to g(x)$, where the convergence is known to be uniform on an interval (a, b) and where

$$f \in C^2(a, b) \quad \text{and} \quad f_n \in C^2(a, b), \quad n = 1, 2, \ldots,$$

$$g \in C^1(a, b) \quad \text{and} \quad g_n \in C^1(a, b), \quad n = 1, 2, \ldots.$$

The solution u_n corresponding to Cauchy data f_n, g_n for the homogeneous wave equation in E^2 is

$$u_n(x, y) = \tfrac{1}{2}[f_n(x+y) + f_n(x-y)] + \tfrac{1}{2}\int_{x-y}^{x+y} g_n(s)\, ds. \tag{5.1.1}$$

Taking the limit of both sides,

$$\lim_{n \to \infty} u_n = \tfrac{1}{2}\left(\lim_{n \to \infty} f_n(x+y) + \lim_{n \to \infty} f_n(x-y)\right) + \tfrac{1}{2}\lim_{n \to \infty}\int_{x-y}^{x+y} g_n(s)\, ds \tag{5.1.2}$$

$$= \tfrac{1}{2}\left(f(x+y) + f(x-y)\right) + \tfrac{1}{2}\int_{x-y}^{x+y} g(s)\, ds, \tag{5.1.3}$$

and we thus see that the Cauchy problem for the wave equation in E^2 is continuously dependent on the data in the sense of Hadamard. We have been able to exchange the order of taking limits and integrals for the sequence $\{g_n\}$ by virtue of the fact that $\{g_n\}$ is uniformly convergent over the interval of integration. Of course, this is true even if this integration is intended in the sense of Riemann. This too will call attention to the classical nature of continuous dependence in the sense it was originally formulated by Hadamard.

2 EXPECTATIONS

It takes guidance and direction to develop sound judgments as to just how far one can anticipate that the results we have so far studied will be acceptable as a general guide. If one *consciously* looks for what could be *anticipated* from

the examples studied thus far, we hope it would prevent leaping to con-
clusions and would counteract any impression that what is still merely
anticipation is, in actuality, established mathematical fact.

We may now ask, from the examples of boundary-value problems given in
Chapter 2, what boundary-value problems do we anticipate will be well
posed for second-order equations of hyperbolic type in E^2?

EXERCISE 1 Prove that, except for the Dirichlet problem, all the problems
treated in Chapter 2 have solutions that are continuously dependent on the
data in the sense of Hadamard.

First, we should expect that, as shown in Fig. 23a, the Cauchy problem
expressed on a sufficiently smooth segment of a noncharacteristic, where u
and its normal derivative are prescribed, is well posed in the forward or
retrograde characteristic "triangle."

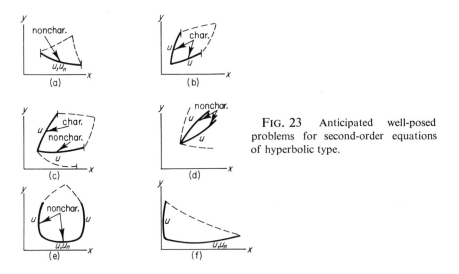

FIG. 23   Anticipated   well-posed
problems for second-order equations
of hyperbolic type.

Second, as shown in Fig. 23b, we should expect that the characteristic
problem, with primitive data only prescribed on intersecting segments of
sufficiently smooth characteristics, is well posed in the characteristic rectangle.

As shown in Fig. 23c, we should expect that the first mixed problem, with
primitive data prescribed on intersecting segments of a characteristic and a
noncharacteristic, with sufficient smoothness, is well posed if both segments

lie in the angle between characteristics. A similar anticipation can be drawn from the Goursat problem as shown in Fig. 23d.

EXERCISE 2 Show that, from the first two anticipations, if a complete character-istic ray lies between the characteristic and noncharacteristic segment, the solution is only unique up to the prescription of another function.

As shown in Fig. 23e, if Cauchy data are prescribed on a noncharacteristic segment and primitive data are prescribed on two segments intersecting the endpoints (all segments sufficiently smooth) and containing a characteristic in the angles, then this problem should be well posed. The presumably inviscid flow pattern inside the expanding side of a converging–diverging nozzle presents such a problem if the streamfunction equation for time-independent, axially symmetric flow is used (see exercises in Chapter 1), and ψ, ψ_z are given on some transverse line downstream of the sonic line (see Fig. 24). This, being in the supersonic region, prescribes Cauchy data on a non-characteristic segment while the walls, where $\psi = 0$, present the intersecting segments. Since the equation is nonlinear, it will have to be checked carefully that any given values of ψ and ψ_z do not make the segment on which they are specified into a characteristic segment (see Exercise 20, Chapter 1). One can see that the anticipation that this problem is well posed is related to the above three such anticipations.

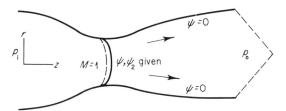

FIG. 24   Converging–diverging nozzle flow, $p_i \gg p_o$.

As shown in Fig. 23f, if Cauchy data are prescribed on a noncharacteristic ray and primitive data are prescribed on one noncharacteristic segment intersecting it at its single end, then from the above anticipation or from the first two we should expect the problem to be well posed in the "triangle" bounded above by a characteristic intersecting both data segments. Trans-mission lines, where a voltage signal v is placed at a point $x = 0$ for a time (t_0, t_1) on a line which is quiet at $t = t_0$ so that $v(x, t_0) = v_t(x, t_0) = 0$, exhibit such a problem. In this connection the student should reexamine the results of Exercise 11, Chapter 1.

In the above, of course, we take it that the student understands he is not being given altogether complete mathematical prescriptions and, in particular, that he knows it is his responsibility to see what conditions must be specified on the data or the functions appearing in the differential equation in order that solutions exist. He would be well advised to ask himself how far these ideas will extend to higher-dimensional problems. In one of our later chapters it will be shown that the idea of a Cauchy problem for the homogeneous wave equation in E^n is still valid where data is now given on a region of a surface, say on a disk in a coordinate plane $z = 0$. In another chapter the matters of most of the above anticipations will be pursued relative to the nonhomogeneous wave equation in E^2.

Turning now to the problems of elliptic type as outlined in Chapter 3, we again examine what we can reasonably anticipate from the examples given.

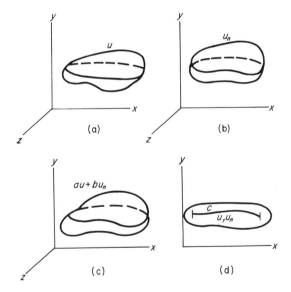

FIG. 25 Well-posed problems for second-order equations of elliptic type.

One should anticipate as shown in Fig. 25a–c that the Dirichlet, Neumann, and mixed problems (specification here of some linear combination of u and u_n), interior and exterior, of second-order equations of elliptic type in any number of dimensions are well posed. We will certainly not attempt to prove so much here, but almost as much is now known,[†] as of very recent date. Virtually the only restriction will be that the region must be one for which the divergence theorem is valid (a Green region) and this is certainly a modest restriction. Even this can sometimes be dispensed with.

With respect to the Cauchy problem, it must be said that it is not continuously dependent on the data in the sense originally formulated by Hadamard. This was the point of his formulation, for he immediately presented an example of uniformly convergent data functions which failed to present a sequence of solutions converging uniformly to the solution.

Let us, however, examine the stability problem for the solution (3.6.1) [or (3.6.2)]. Let f be perturbed by a function δ, and g by a function η, both defined on (a, b), and let ψ be displaced by a function v. Then subtracting the Laplace equation, presumed valid for ψ and $\psi + v$, we have

$$v_{xx} + v_{yy} = 0 \tag{5.2.1}$$

and we have

$$v(x, 0) = \delta(x), \qquad v_y(x, 0) = \eta(x). \tag{5.2.2}$$

The unique solution of this problem is

$$v(x, y) = \operatorname{Re} \delta(x + iy) + \operatorname{Re} \int_0^y \eta(x + it)\, dt. \tag{5.2.3}$$

Of course, the displacement may not exist—may even "be infinite"—unless δ and η are carefully prescribed to have usable properties. We give an example (which will here serve to replace the more difficult Hadamard example) to demonstrate the possible disastrous implications with respect to stability. Let $g(x)$ be unperturbed so that $\eta(x) = 0$ on $(-1, 1)$, and let $f(x)$ be perturbed by an amount

$$\delta(x) = \varepsilon/(x^2 + \Delta^2). \tag{5.2.4}$$

Both perturbations are analytic on the real interval $(-1, 1)$ so they do not imply any violations of conditions we found to be necessary in Chapter 3 for the initial functions, but we have

$$\max_{x \in (-1, 1)} \frac{\varepsilon}{x^2 + \Delta^2} = \frac{\varepsilon}{\Delta^2}, \tag{5.2.5}$$

---

† With odd-shaped regions, care must be exercised, however. In E^3 a logarithmic horn, or "apple shape" can cause difficulty.

which, for any given choice of Δ, can be made arbitrarily small by selection of ε. However, the resulting displacement is

$$v(x,y) = \operatorname{Re} \frac{\varepsilon}{(x+iy)^2 + \Delta^2}, \qquad (5.2.6)$$

and we see that it is unbounded in a neighborhood of the point $(0, \Delta) \in E^2$, which can be made arbitrarily close to the data line by choice of Δ.

We can say, however, that (3.6.1) [or (3.6.2)] *is stable with respect to data perturbations that are in a class of functions which are analytic in appropriate extended regions in the complex plane.* This avoids the difficulty, of course, with the perturbation (5.2.4). There is a curious sense of direction with respect to these stability considerations: If both δ and η have singularities at the complex point $x + i\Delta$, the solution may still exist at $(x, \Delta) \in E^2$ if

$$v(x, \Delta) = \operatorname{Re} \delta(x+i\Delta) + \int_0^\Delta \operatorname{Re} \eta(x+it)\, dt \qquad (5.2.7)$$

has no singularity because, clearly, there may be some cancelation of singularities, depending on signs of δ and η. But then the displacement

$$v(x, -\Delta) = \operatorname{Re} \delta(x+i\Delta) - \int_0^\Delta \operatorname{Re} \eta(x+it)\, dt \qquad (5.2.8)$$

at $(x, -\Delta)$ does not exist; for if it did, then adding δ would have no singularity at $(x+i\Delta)$ contrary to assumption.

In the sense of stability with respect to classes of perturbations analytic in given regions of the complex plane, not in the sense formulated by Hadamard, one should expect the Cauchy problem on a segment of an analytic curve with analytic data to be well posed. Solutions will exist in some neighborhood of points of the data segment and be stable there, but these neighborhoods must be expected to depend on the region in the complex plane where the perturbation functions are analytic.

Finally, we examine what we can anticipate from the material of Chapter 4 concerning well-posed problems for parabolic type.

One should anticipate purely from the examples in Chapter 4 with regard to the heat equation that the Cauchy problem, where primitive data is prescribed on a doubly infinite characteristic curve (as shown in Fig. 26b), or on a half-infinite characteristic ray and on an intersecting noncharacteristic segment in a "positive timelike direction" (as shown in Fig. 26c) or on a characteristic segment and on two noncharacteristic segments "in a positive timelike direction" intersecting its ends (as shown in Fig. 26a) to be well posed. It is to be noted that we have clearly demonstrated in Chapter 4 that,

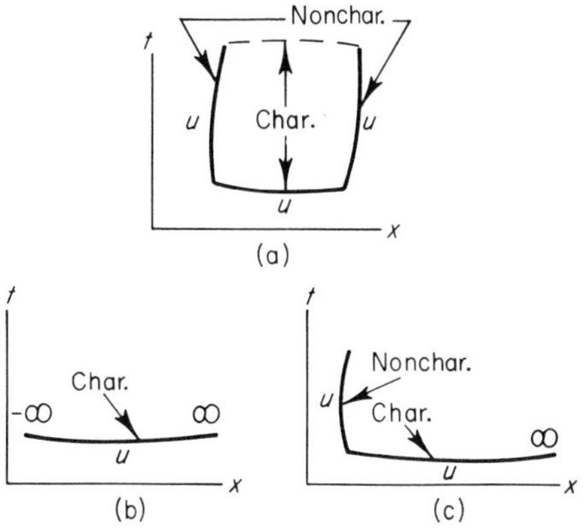

FIG. 26 Well-posed problems for second-order equations of parabolic type.

for the heat equation, the slab problem with t replaced by $-t$ is asymptotically unstable with respect to perturbations of the initial data, so such a problem is not continuous with respect to its data in the sense of Hadamard, and no restriction of the class of perturbations allowed will be helpful in this respect. Obviously, then, the student must be on his guard in attempting to use these anticipations as to what constitutes our vague notion above of "a positive timelike direction" for his particular equation. As elsewhere here, nothing will substitute for some close analysis of his particular problem. Knowing what to look for should, however, be helpful.

But unlike the Laplace and wave equation, it turns out that the heat equation is not entirely typical of its type. We have tried to anticipate this in Chapter 4 by presenting a solution of a Dirichlet problem for a simple equation of parabolic type.

Before we pursue the specification, however, of further problems of parabolic type—or, really, of elliptic–parabolic type—we pause to settle the ambiguities that may have been raised so far about what really is meant by "continuous dependence." The modern interpretation of this concept is very direct. The modern mathematician intends simply to say that a boundary-value problem is continuously dependent on its data if, in a modern functional analysis sense, the solution function (vector) is a continuous function of the data function (vector). We will find in Part IV the place to introduce many of the basic concepts of functional analysis, and, in particular, the place to find how this kind of continuity is expressed. For it, one needs a way of measuring the size of functions, a "norm," which concept is a generalization of that of an absolute value of a real variable. The norm of a function f is denoted $\|f\|$, and it is a real number. We will find that for linear problems, a solution u is a continuous function of boundary data f if there exists a real number K such that

$$\|u\| \leqslant K\|f\|, \tag{5.2.9}$$

and this then becomes the meaning of continuous dependence for linear problems.

3 BOUNDARY-VALUE PROBLEMS FOR EQUATIONS OF ELLIPTIC–PARABOLIC TYPE

A matrix (a^{ij}) is said to be positive definite if for all real λ_i, λ_j, not all zero,

$$a^{ij}\lambda_i\lambda_j > 0 \tag{5.3.1}$$

and positive semidefinite if

$$a^{ij}\lambda_i\lambda_j \geqslant 0. \tag{5.3.2}$$

It is this distinction, when used on the matrix of coefficients of a linear second-order partial differential operator, that distinguishes operators of elliptic type from those of elliptic–parabolic type *in any number of dimensions*. The class of operators which are not of elliptic type but are of elliptic–parabolic type, even in two dimensions, is not necessarily typified by the heat equation, and this is the nature of our contention in the section above. Of course, in (5.3.1) and (5.3.2) we have assumed that repeated indices i, j would be summed from 1 to n. This (Einstein) summation convention will be adopted in the sequel when it is useful.

Consider a region $A \subset E^n$ with boundary Σ. The general second-order linear operator

$$Lu = a^{ij}(x)u_{x_ix_j} + b^i(x)u_{x_i} + c(x)u \tag{5.3.3}$$

is said to be of elliptic–parabolic type in A if for every $x \in A \cup \Sigma$

$$(a^{ij}(x)) \quad \text{is positive semidefinite} \tag{5.3.4}$$

and

$$a^{ij}(x) \in C^2(A \cup \Sigma), \quad b^i(x) \in C^1(A \cup \Sigma), \quad c(x) \in C(A \cup \Sigma). \tag{5.3.5}$$

It will be found also that in order to prove a maximum and other important principles needed to treat of boundary-value problems that it will be required that

$$c(x) < 0. \tag{5.3.6}$$

Where n_i, n_j are direction cosines of the inner normal to points of the boundary Σ, we divide the boundary into three disjoint sets of points as follows (see Fig. 27):

$$\Sigma^3 = \{x | (x \in \Sigma) \wedge a^{ij}(x) n_i n_j \neq 0\}$$

$$\Sigma^2 = \{x | (x \in \Sigma - \Sigma^3) \wedge b(x) = (b^i(x) - a^{ij}_{x_j}(x)) n_i < 0\} \qquad (5.3.7)$$

$$\Sigma^1 = \{x | (x \in \Sigma - \Sigma^3) \wedge b(x) = (b^i(x) - a^{ij}_{x_j}(x)) n_i \geq 0\}.$$

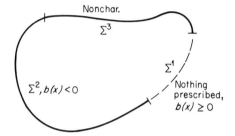

FIG. 27   Well-posed boundary-value problems for equations of elliptic–parabolic type.

The set Σ^3 is the noncharacteristic portion of the boundary, and the discriminator function $b(x)$ will be used to delineate on which portions of the boundary data may or may not be specified. If the operator (5.3.3) is of elliptic type, then evidently the entire boundary is $\Sigma^3 = \Sigma$.

EXERCISE 3 For the Dirichlet problem for the Fichera equation, which is of parabolic type, solved in Chapter 4 show that the entire boundary of the square is Σ^3 with the exception of four isolated points.

Of course, A should be such that the normal is defined except on a set of measure zero.[†] It will later become evident that we will need A to be a Green region (one where the divergence theorem is valid), but this is not important for now and represents only a minor restriction.

We consider the boundary-value problem

$$Lu = g(x) \qquad \text{for} \quad x \in A \qquad (5.3.8)$$

$$u = f(x) \qquad \text{for} \quad x \in \Sigma^2 \cup \Sigma^3.$$

EXERCISE 4 Show that the Dirichlet problem for the modified Laplace equation in E^n, $u_{x_i x_j} - u = 0$, is included in the above.

[†] For now, the student may think of this as a denumerable set of points.

EXERCISE 5 Show that the slab problem for the modified heat equation, $u_{xx} - u_t - u = 0$, is included in the above.

We will eventually (Part IV) undertake to prove that the large class of boundary-value problems described by (5.3.8) is well-posed, but this cannot be in the primitive sense originally formulated by Hadamard. Existence will be in $\mathscr{L}^p$-weak sense, and continuous dependence will be in the usual modern setting described in (5.2.9) above. Separate discussions of the uniqueness of regular and weak solutions will be given, and some inequalities that are useful for computing error bounds (see Chapter 3, Section 5) will be obtained in the process. The original works of Fichera on this class of problems include the Neumann and mixed problems for the modified Laplace equation, but we have chosen to restrict the amount of this material to be included here, and thus consider only the Dirichlet problems (5.3.8).

Since these matters will not be taken up until the last part of this text, we now attempt to describe the current milieu of generalizations of the sense of existence, ending in a description of what is meant by $\mathscr{L}^p$-weak solutions.

4 EXISTENCE AS THE LIMIT OF REGULAR SOLUTIONS

One sense of generalization of the concept of existence of solutions of boundary-value problems, perhaps the earliest, is quite easily understood in terms of solutions[†] to the Cauchy problem for the homogeneous wave equation in E^2. Let us consider this problem for the case where it is known only that the data functions

$$f, g \in C(a, b), \tag{5.4.1}$$

and, perhaps, where it is even known that derivatives of these functions do not exist at some points of (a, b). In that case it is even known that no regular solution exists to the Cauchy problem. However, the well-known Stone–Weierstrass theorem tells us that there exist sequences of (Bernstein) polynomials, $\{f_n\} \to f$, $\{g_n\} \to g$, which converge to f and g uniformly on (a, b). We now return to Eq. (5.1.1) and the argument following it. To each pair f_n and g_n (since they are polynomials and thus at least C^2 and C^1, respectively) corresponds a regular solution u_n; and, taking the limit of the sequence $\{u_n\}$, because of the uniform convergence of $\{f_n\}$ and $\{g_n\}$, we find that this limit can be written in the form (5.1.3) (even with respect to the strongest

---

[†] In Chapter 1 we proved that the regular solution of this problem is unique, but when we begin to speak of possible generalizations of the sense of existence, the fact of uniqueness is by no means immediately clear.

possible sense of integration). But this is precisely the solution we would quote if indeed it were true that $f \in C^2$ and $g \in C^1$. There is nothing wrong with the function (5.1.3) per se—the integral clearly exists, and it even satisfies the stated boundary conditions. But what is its relation to the differential equation? Certainly it cannot simply be substituted term by term in the differential equation because these terms require the taking of two derivatives which this expression may not, or does not, have. Yet it is certainly an expression closely related to this problem and, moreover, closely related to any physical phenomena it may be designed to describe. It is, of course, simply the limit of a sequence of regular solutions, and this is our very concrete (and very evident) generalization of the concept of existence.

It is seen above that in this sense a solution exists to the Cauchy problem for the homogeneous wave equation in E^2 with continuous data, and uniqueness is even evident. "Completion by ideal vectors" has been a common technique even in the early developments of Hilbert space theory, and this idea is very closely related. There a pre-Hilbert space, one having all the necessary properties of a Hilbert space except completeness—a linear vector space with a bilinear form yielding a positive-definite quadratic form (norm)—is constructed out of the (perhaps regular) solutions of a problem. Then, in imitation of the Cantor formulation of the real number system from the rationals, equivalence classes of Cauchy sequences of space elements are formed under a relation which identifies those sequences whose differences converge to the zero element of the space. The natural addition and multiplication by scalars is formed on these classes by reflection from those in the pre-Hilbert space, and the natural norm is assumed. The new system (which has the pre-Hilbert space embedded in it) is found, of course, to be complete. That is to say, the equivalence classes now form a normed space with inner product, complete in respect to the norm, and it is the elements of this space that are taken to be solutions.

Yet, the student should once again note that the declared solution cannot be substituted into the differential equation and be found to satisfy it. Undoubtedly it will come as a shock to many engineers and physicists that modern mathematicians rarely mean that a function "satisfies" a differential equation when they say it is a solution of a boundary-value problem related to it.

5 THE IMPULSE PROBLEM AS A PROTOTYPE OF A SOLUTION IN TERMS OF DISTRIBUTIONS

Our presentation of the solution to the classical "impulse problem" for the diffusion or heat equation presented in Section 5 of Chapter 4 already clearly indicated, without specially naming it as such, a generalization of the sense of existence in order to be able to discuss the problem of uniqueness. The limit of a sequence of initial functions, ever sharpening and converging point for point to the impulse conditions required for $t = 0$, was said to be the intrinsic meaning of a problem modeling the deposition of a finite mass of diffusing substance at a single point by calling for an infinite concentration there. This limit, then, is a solution in much the same generalized sense given above except for one major difference. *It cannot be said that the sequence of initial functions converges uniformly to the initial impulse data.*

In fact, the initial impulse data can be said to be a function by only the kindest stretch of the sense we usually impose on such things, by using the real number system extended to include ∞. This is not done, for good reason usually, in discussing the function defined by $1/x$, for example, in a neighborhood of the origin. The impulse needed is at best a most peculiar function with no usable properties and is obviously better thought of as being the sequence of functions converging pointwise to its values at all points except $x = 0$, in a neighborhood of which it is unbounded. With a little more precision of statement, it could be thought of as the class of function sequences with this property.[†] These are the data. The solution in this problem does not give a similar level of difficulty in expressing it as a function, but the solution of some other similar problems do.

Growing out of similar problems and similar difficulties has been the intensively developed modern theory of distributions or theory of generalized functions which has broad areas of application that sometimes seem far removed from the study of boundary-value problems. The original impetus, however, came from a propensity of modern physicists, apparently born of necessity and essentially starting from the work of Dirac on the so-called "delta function," to treat of equations where impulses occur in the differential equation. We will see in the following material that regular solutions of homogeneous linear differential equations corresponding to nonzero boundary conditions usually can be thought of as solutions of nonhomogeneous differential equations corresponding to zero (homogeneous) boundary conditions. Therefore, the example in Chapter 4, Section 5 can be taken as an example of an equation containing an impulse. The theory of distributions is available in a number of more or less elementary sources and in many erudite sources, and we will not attempt to treat it here. Apparently the relation between solutions in this sense and in the $\mathscr{L}^p$-weak sense we will now begin to develop has not been carefully delineated, but the relation is clearly a close one.

---

† But then the proof of Section 5, Chapter 4, for uniqueness is not complete.

6 THE GREEN IDENTITIES

Several identities for functions having two continuous derivatives are arrived at by direct application of the divergence theorem (sometimes called the Gauss theorem) and are closely enough related to the divergence theorem that authors often confuse them without being thus held accountable for an error. In his monumental work on potential theory, Kellogg [24] presented three Green identities, two of which we now present, leaving the third to a later chapter.

The divergence theorem concerns a region[†] $A \subset E^n$ with topological boundary δA, and a function $a: A \cup \delta A \to E^n$ such that

$$a \in C^1(A) \qquad \text{and} \qquad a \in C(A \cup \delta A) \qquad (5.6.1)$$

If $x = (x_1, \ldots, x_n) \in A$, we write $a_i(x) = a_i(x_1, \ldots, x_n)$, $i = 1, \ldots, n$, for each of the real component functions a_i, and put

$$\operatorname{div} a = \sum_{i=1}^{n} \partial a_i / \partial x_i. \qquad (5.6.2)$$

If $\bar{n}$ is the outer normal on the boundary of A, the divergence theorem then purports to state that

$$\int_A (\operatorname{div} a) \, dA = \int_{\delta A} (a, n) \, ds \qquad (5.6.3)$$

where

$$(a, n) = \sum_{i=1}^{n} a_i n_i \qquad (5.6.4)$$

and the n_i are direction cosines of $\bar{n}$. If we use the inner normal, the sign of one integral in (5.6.3) is altered. The theorem, or some form closely related to it, is indispensible whenever it is required to convert a volume integral into a surface integral. If A is a rectangle in E^2, this is simply the expression of the Fubini theorem [22, 41] which states that an area integral can be written as an iterated integral (first in x, then in y), and a similar statement can be made in higher dimensions.[‡] It was already used in this form in Chapters 2, 3, and 4 to prove uniqueness in E^2 for the vibrating string problem, the Dirichlet–Neumann mixed problem, and the slab problem; utilizing the divergence theorem, these proofs can now be extended to rather arbitrary regions in E^n.

---

† An open, connected set, not necessarily simply connected.

‡ For a much more sophisticated context in which this theorem is often stated, the student may wish to read about exterior differential forms.

This extension is to the class of regions for which the divergence theorem is valid, an extremely large class now called "Green regions."

EXERCISE 6 Show that the divergence theorem is valid for rectangles in E^2 and rectangular prisms in E^n.

EXERCISE 7 Show that the divergence theorem is valid (a) on triangles, (b) on trapezoids, (c) on any region which is a finite union of trapezoids.

EXERCISE 8 Show that the divergence theorem is valid (a) on triangular prisms[†] in E^3, (b) on any region which is a finite union of triangular prisms.

It is shown in a number of elementary sources that the divergence theorem is valid in any region composed of a limit of regions of the class treated in the exercises above. Whether or not other regions, such as ones having a boundary point like $x = 0$ in $\sin(1/x)$, are Green regions is another question. We are not aware of any necessary and sufficient conditions yet available. The study of this question is one deeply ingrained in the study of algebraic topology. It is, in fact, essentially the study that prompted the beginnings of the field in an early form called *analysis situs*. In algebraic topology the concept of a "simplex" is introduced in E^n, and it replaces the role of the triangle in E^2 or that of the tetrahedron in E^3. Like the pursuit of foundations questions for an analyst generally, one who is interested in partial differential equations should undertake such studies only with the full realization that they will take him far afield from this pursuit. A student, however, should here recognize that algebraic topology is one of the areas he truly needs to study for a complete underpinning of his basic concepts. From some points of view, however, that we regard as thoroughly valid, a thorough study of partial differential equations requires a thorough study of all of science, and thus there is always, it must be admitted, some aura of mystery.

But we have already indicated an extremely large class of Green regions, so let us work here in this context, conscious, of course, that this class may be extended. Let $u: A \to E^1$ and $u \in C^1(A)$, then we put

$$\operatorname{grad} u = (u_{x_i}) \tag{5.6.5}$$

where it is understood that $\operatorname{grad} u$ is vector (or column matrix) valued.

---

† These are cylindrical surfaces following a triangle as directrix, cut to finite length by parallel planes.

EXERCISE 9 Show that for $u, v \in C^1(A)$

$$\operatorname{div}(u \operatorname{grad} v) = (\operatorname{grad} u, \operatorname{grad} v) + u \Delta v.$$

We now see that if A is a Green region in E^n, for every $u, v \in C^2(A)$, $C^1(A \cup \delta A)$, from (5.6.3) with

$$a = u \operatorname{grad} v$$

we have

$$\int_A u \Delta v \, dA + \int_A (\operatorname{grad} u, \operatorname{grad} v) \, dA = \int_{\delta A} (\bar{n}, u \operatorname{grad} v) \, ds$$

$$= \int_{\delta A} u(\bar{n}, \operatorname{grad} v) \, ds$$

$$= \int_{\delta A} u(\partial v / \partial n) \, ds. \qquad (5.6.6)$$

The equation (5.6.6), since it is valid for all functions u and v in the large class stated, is called an identity, the first Green identity according to Kellogg. Interchanging roles of u and v in (5.6.6) [i.e., putting $a = v \operatorname{grad} u$ in (5.6.3)] and subtracting we obtain the second Green identity for every $u, v \in C^2(A)$, $C^1(A \cup \delta A)$,

$$\int_A (u \Delta v - v \Delta u) \, dA = \int_{\delta A} (u \, \partial v / \partial n - v \, \partial u / \partial n) \, ds. \qquad (5.6.7)$$

The breadth of consequences deriving from (5.6.7) and its generalizations are breathtaking. Yet if one wishes to be flippant about the questions in algebraic topology raised above, it could easily be said that it is simply the Fubini theorem or at best a matter of integration by parts!

7 THE GENERALIZED GREEN IDENTITY

The general second-order linear partial differential operator on a region $A \subset E^n$ can be written as

$$Lu = A^{ij} u_{x_i x_j} + B^i u_{x_i} + Cu \qquad (5.7.1)$$

which, of course, is to be applied to functions $u \in C^2(A)$. Here we presume that, in fact, for every $i, j = 1, \ldots, n$,

$$A^{ij} \in C^2(D), \qquad B^i \in C^1(D), \qquad C \in C(D). \qquad (5.7.2)$$

89

Lagrange defined the adjoint L^* of L as follows [for every $v \in C^2(D)$]

$$L^*v = (A^{ij}v)_{x_i x_j} - (B^i v)_{x_i} + Cv, \tag{5.7.3}$$

and then presented the following identity for functions $u, v \in C^2(A)$, $C^1(A \cup \delta A)$, written here in two forms,

$$
\begin{aligned}
vLu - uL^*v &= \sum_{i=1}^{n} \frac{\partial}{\partial x_i} \left\{ \sum_{j=1}^{n} vA^{ij}u_{x_j} - u(A^{ij}v)_{x_j} + B_i uv \right\} \\
&= \sum_{i=1}^{n} \frac{\partial}{\partial x_i} \left\{ \sum_{j=1}^{n} A^{ij}(v\partial u/\partial x_j - u\partial v/\partial x_j) + uv\left(B^i - \sum_{j=1}^{n} (A^{ij})_{x_j} \right) \right\} \\
&= \sum_{i=1}^{n} \partial/\partial x_i \{P_i\} = \operatorname{div} P
\end{aligned} \tag{5.7.4}
$$

where $P = (P_1, P_2, \ldots, P_n)$ has the appropriate definition indicated by the equalities. The derivation of (5.7.4) might offer some difficulty if, as Lagrange, we were faced with the task of defining L^* appropriately so that the form $vLu - uL^*v$ could be presented as a divergence of some vector P, or even if, knowing this definition of L^*, we were asked to derive a vector P so that this form be presented as $\operatorname{div} P$. But, having the right side of (5.7.4) in hand, it only requires the carrying out of certain derivative manipulations to verify it. We have

$$
\sum_{i=1}^{n} \partial P/\partial x_i = [vA^{ij}u_{x_i x_j} + vB^i u_{x_i} + Cuv] - [u(A^{ij}v)_{x_i x_j} - u(B_i v)_x + Cuv]
$$
$$
+ [u_{x_j}(A^{ij}v)_{x_i} - u_{x_i}(A^{ij}v)_{x_j}]. \tag{5.7.5}
$$

The last bracket vanishes because the two groups of terms composing it are equal to each other with i and j interchanged while all factors are commutative, $A^{ij} = A^{ji}$ [for $u, v \in C^2(A)$, see (5.7.1)], and the two groups of terms are subtracted. The other two brackets are vLu and uL^*v just as stated in (5.7.4).

At this point one should carefully note the placing of parentheses in (5.7.3) and the alteration of sign in passing from (5.7.1) to (5.7.3) as well as the close relation of (5.7.4) to (5.6.7). Lagrange originally designed the adjoint (5.7.3) and the generalized Green identity (5.7.4) specifically for purposes of work like that to be treated here in partial differential equations. Since that time, the two concepts have been so successful that they have been adapted to other purposes far beyond the original role intended. Yet wherever such phrases are used one can be assured that if he looks deeply the relation will be discovered.

8 $\mathscr{L}^p$-WEAK SOLUTIONS

Let A be a Green region. Clearly, for f sufficiently smooth a function $v\colon A \cup \delta A \to E^1$ exists such that $v \in C^2(A \cup \delta A)$ and such that

$$v = f \quad \text{in} \quad \delta A. \tag{5.8.1}$$

We have not asked that v satisfy any differential equation, only that it satisfy a boundary condition. In fact, having chosen v so that it satisfies (5.8.1), we denote Δv by g. It is then evident that the Dirichlet problem for the Laplace equation

$$\Delta w = 0 \quad \text{in} \quad A, \qquad w = f \quad \text{in} \quad \delta A \tag{5.8.2}$$

can be written as a Dirichlet problem for the Poisson equation (nonhomogeneous Laplace equation) by putting

$$u = w - v, \tag{5.8.3}$$

from which

$$\Delta u = g \quad \text{in} \quad A, \qquad u = 0 \quad \text{in} \quad \delta A. \tag{5.8.4}$$

Next we notice that there are many functions for which

$$v \in C^2(A), \qquad v \in C^1(A \cup \delta A), \qquad \text{and} \qquad v = 0 \quad \text{in} \quad \delta A. \tag{5.8.5}$$

They are sometimes called "test functions" or even "functions with compact support on A of order 2." Again, we do not ask that functions v should satisfy any differential equation, only that they should satisfy a boundary condition. It can be shown, in fact, that the class of functions that satisfy (5.8.5) is dense in the space of $\mathscr{L}^p$ functions which vanish on δA. These functions are those for which the pth power of the absolute value is integrable in the Lebesque sense.

EXERCISE 10 Give examples of functions satisfying (5.8.5) for cases where A is a rectangle or a circle.

Consider now the presumed regular solution u of the boundary-value problem (5.8.4) and the class of all functions v satisfying (5.8.5) substituted into the second Green identity (5.6.7). Since $u = v = 0$ on δA, the boundary integrals vanish, and we then have that

$$\int_A (u\,\Delta v - v\,\Delta u)\, dA = 0 \tag{5.8.6a}$$

or

$$\int_A u \, \Delta v \, dA = \int_A v g \, dA \qquad (5.8.6b)$$

If there exists a function $u \in \mathscr{L}^p$ (not necessarily C^2) such that (5.8.6b) is satisfied for every v satisfying (5.8.5), then u is said to be an $\mathscr{L}^p$-weak solution. To try to paraphrase one's feelings: It may be said then that such a function has all those properties of being a solution of (5.8.4) which are relevant to the basic integral identities of mathematical physics, and these have come to play such a central role that we may in some cases be willing to ignore all else.

If one looks at the derivations of differential equation models for physical problems from variational principles (the Hamilton principle), he finds that equations like (5.8.6b) actually occur as a step in the derivations. If besides belonging to $\mathscr{L}^p(A)$, we find that a weak solution u actually belongs to $C^2(A)$, then (5.8.6a) applies and we find then that

$$\int_A v(\Delta u - g) \, dA = 0 \qquad (5.8.7)$$

for every v satisfying (5.8.5). But since this class of functions is dense in the functions belonging to $\mathscr{L}^p$ which vanish in δA, we have that

$$\Delta u - g = 0 \quad \text{in} \quad A. \qquad (5.8.8)$$

Lagrange was willing in his derivations to make this kind of regularity assumption on functions u that are to be solutions of problems with physical meaning and thus came to differential equations like (5.8.8) from his variational methods.

To see the broad range of applicability of the concept of $\mathscr{L}^p$-weak solutions, consider now the generalized Green identity (5.7.4) and consider a boundary-value problem

$$Lu - 0 \quad \text{in} \quad A, \qquad u = f \quad \text{in} \quad \delta A,$$

or, since we have seen that it is equivalent, the boundary-value problem

$$Lu = g \quad \text{in} \quad A, \qquad u = 0 \quad \text{in} \quad \delta A. \qquad (5.8.9)$$

Then from (5.7.4), for all $v \in C^2(A)$ such that $v = 0$ in δA, we have that

$$\int_A (vLu - uL^*v) \, dA = 0$$

or

$$\int_A (uL^*v)\, dA = \int_A vg\, dA. \tag{5.8.10}$$

Again, if there exists a function $u \in \mathscr{L}^p(A)$ such that (5.8.10) is satisfied for all $v \in C^2(A)$ such that $v = 0$ in δA, it is said to be an $\mathscr{L}^p$-weak solution of (5.8.9). As we have now often stressed, we will eventually prove the existence of $\mathscr{L}^p$-weak solutions to the class of problems of elliptic–parabolic type outlined in Section 3. Because of the usefulness of regular solutions, one often proves the existence of $\mathscr{L}^p$-weak solutions for a large class of boundary-value problems and then proceeds to undertake difficult arguments in special cases to show that a solution can be regularized. This kind of technique will not be treated here.

9 PROSPECTUS

The student has now been introduced to the three major classes of the sense of generalized solutions; namely, completion by ideal vectors, distributions, and $\mathscr{L}^p$-weak solutions. Others do occur; for example, functions that minimize a quadratic form for which the solution could be shown to satisfy a differential equation (if they were regular) are sometimes said to be energy-weak solutions of that differential equation. Also, some variations in terms used may occur among authors, and even from one work of an author to another work there usually will occur a difference in the class of test functions v utilized. For nonlinear equations, there would still seem to be major disagreements as to what should be called a weak solution, while for linear equations the variations mentioned certainly do not represent gaps.

The student has seen an example in Chapter 4, that of (viscous) pipe flow, where mean-square asymptotic stability could be accepted as a weakened sense of uniqueness, giving uniqueness, in fact, up to short-lived transients. This ploy—in problems, for example, involving tubes with compliant walls— gives leeway for many solutions to exist and makes it possible to consider regular solutions,[†] which are physically and computationally far easier to utilize. *Perhaps* the really misleading step of the past has been to consider time-independent problems. *Perhaps* we should always in physics consider time-dependent problems where a damping or other mechanism exists to produce an eventual steady state, and only this state (not necessarily independent of time) should be expected to be unique.

---

† Without undertaking "regularization."

10 THE TRICOMI PROBLEM

We hasten to append to this outline a brief discussion of a problem that we did not find a place for in the above discussions. It will be recalled that in Exercise 4 of Chapter 1, the student was asked to find the characteristics of the Tricomi equation

$$u_{yy} + yu_{xx} = 0. \tag{5.10.1}$$

These were two families of semicubical parabolas in the lower half-plane, and the equation is of elliptic type for $y > 0$, parabolic type on the line $y = 0$. This equation is said to be of mixed elliptic–hyperbolic type and is thought to be as typical of this class as the Laplace equation is of elliptic type and the wave equation is of hyperbolic type. Tricomi's mixed problem for the mixed-type equation (5.10.1) is pictured in Fig. 28. Primitive data u are given on a semicircular arc, the ends of which lay at the origin and on a point of the x axis to the right of it, and on a characteristic segment with one end at the origin. Data are not specified on the characteristic segment which would complete the boundary. This is allowed to become what we have called a "natural boundary" for the region of determination.

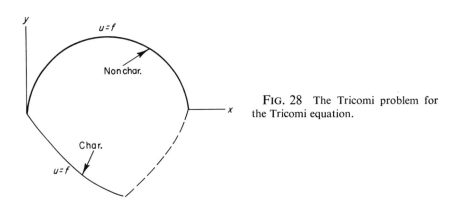

FIG. 28   The Tricomi problem for the Tricomi equation.

Tricomi proved existence and uniqueness by a method of limits, constructing two line segments a distance ε above and below the x axis and giving the same data function $u = f$ on both segments. Taking representation formulas expressing solutions in terms of integrals for the corresponding Dirichlet problem in the upper half-plane and the mixed problem in the lower half-plane, he was able to choose a function f so that upon taking the limit as $\varepsilon \to 0$ in each problem, the second derivatives existed on $y = 0$. Undoubtedly, Tricomi's success with this problem and the general appearance of the result shown in Fig. 28 motivated Fichera's general formulation of boundary-value problems for elliptic–parabolic type. So far no similar work has appeared with respect to elliptic–hyperbolic type. Presumably such a formulation would have to include a condition that the transition (parabolic type) occur only on a line segment or surface.

At the turn of the century a Russian scientist, Chaplygin, had shown that in irrotational, two-dimensional, time-independent gas dynamics it was possible to invert the roles of dependent and independent variables so as to obtain linear equations. With velocity components as independent variables, this is the renowned "hodograph method," and Chaplygin had shown that subsonic flow was approximated by the Laplace equation, supersonic flow by the wave equation, and transonic flow by what is now called the Tricomi equation, in the hodograph plane.

II

Some Classical Results for Nonlinear Equations in Two Independent Variables

In Chapter 6 we will develop to some extent the anticipations we have encouraged in Chapter 5 relative to possible generalizations to hyperbolic type of the purely elementary results concerning well-posed problems for the wave equation (as given in Chapter 2). For this purpose, we treat the existence and uniqueness of both the characteristic and initial-value (Cauchy) problems for the possibly nonlinear equation,

$$u_{xy} = f(x, y, u, u_x, u_y). \tag{II.1}$$

Of course, the operator u_{xy} is the operator $u_{\xi\xi} - u_{nn}$ with coordinates ξ, n being the coordinates x, y rotated through 45°, so (II.1) is simply the non-homogeneous wave equation written in characteristic coordinates. The importance of the simplification contained here has already been emphasized in that (II.1) has a linear principle part. The principle part, in fact, has constant coefficients, and it may be noticed that any second-order equations with constant coefficients will have this form in a region of hyperbolicity when written in characteristic coordinates.

Our method for establishing existence and uniqueness will be that of iteration on an equivalent integral equation in a manner precisely analogous to the technique introduced by Picard in his *Comptes Rendus* article of 1894 [34] for the initial-value problem

$$dy/dx = f(x, y), \qquad y(x_0) = y_0 \tag{II.2}$$

where f is required to satisfy a Lipschitz condition. This treatment is one which is given for existence and uniqueness in every good elementary text on ordinary differential equations.[†] A major complication that occurs in treating existence and uniqueness for our partial differential equation problems (II.1) does not occur in the initial-value problem for an ordinary differential equation (II.2). In considerations related to (II.1) it is found that we must keep track of convergence and continuity properties for the two sequences of functions $\{u_x^n\}$ and $\{u_y^n\}$ as well as $\{u^n\}$ and, in the end, appropriately to relate the limits of these sequences. This is done concisely using a single set of inequalities in a manner much like that given in [23]. However, the student may find [42] a more readily available reference for material of this nature. The presentation here is more or less taken from classroom notes received from Professor J. B. Diaz when he was at the University of Maryland Institute of Fluid Dynamics and Applied Mathematics.

Our aim in all this will be to develop the Riemann method (Chapter 7), which is a method providing integral representations of solutions to boundary-value problems for equations of the form (II.1) where f is linear in its last three variables, and then to use this method in the elucidation of classical transmission line theory (Chapter 8). The Riemann function, a resolvent (or kernel) function used in the integral representations of the Riemann method, is the solution of a certain characteristic boundary-value problem for an equation of the form (II.1) where f is linear, and existence–uniqueness is immediate by application of the more general theory for (II.1). Even the Lipschitz condition needed here for f is automatically satisfied on any bounded region when f is linear.

We believe the student will find some thrill in seeing the theory he has learned returned to one of the practical purposes for which it was bred in treating transmission line theory (Chapter 8). Moreover, our comments relative to treatments of the electrical properties of the axons of nerve and muscle tissue will give it a modern context and an exciting purpose.

Part II will be concluded with a treatment (Chapter 9) of the Cauchy–Kovalevsky theorem in E^2 with comments relative to its validity in E^n for any $n \geqslant 2$. This theorem may be regarded as a generalization of the results in Chapter 3 concerning the Cauchy problem for the Laplace equation, only such specific and complete results are not attainable. Existence and uniqueness in a neighborhood of the data line of analytic solutions to "analytic initial-value problems" is all that we are able to produce here, and this falls far short, for example, of the complete stability analysis we gave for the Laplace

---

† A particularly facile proof is given in the finely written text by E. A. Coddington [4].

equation initial-value problem in Chapter 5. This is because we will be unable to say how far from the data line a solution exists. The Cauchy–Kovalevsky theorem predates any knowledge of type classification and is thus the earliest general theorem which is independent of type. The emphasis, rather than on type and on the minimization of the required conditions on data functions, is on analyticity and power series. The requirements of analyticity are usually considered too strong and the results, leaving the exact region of determination unspecified (*some* neighborhood of the data line), are considered too weak, but the theorem is nevertheless often useful [15] and is often "a taking-off point for one's thinking" in somewhat the same way that the D'Alembert solution given in Chapters 1 and 2 is in a more specific context. Some authors (see, e.g., [16]) have thought the Cauchy–Kovalevsky theorem important enough as a taking-off point to place it in the very first pages of their texts. We agree on this status but find the proof possibly too intricate and the results too subtle to appeal to a student before he has sampled other general results in the field.

6

Existence and Uniqueness Considerations for the Nonhomogeneous Wave Equation in E^2

1 NOTATION

The student is reminded that solution of the characteristic boundary-value problem for the homogeneous wave equation,

$$u_{xy} = 0, \qquad u(x,0) = \sigma(x), \qquad u(0,y) = \tau(y),$$

is given by

$$u(x,y) = \sigma(x) + \tau(y) - \sigma(0) \qquad (6.1.1)$$

where $\sigma(0) = \tau(0)$. This is the function ψ_0, defined in the characteristic triangle shown in Fig. 29 and used as the initial iterate in the Picard iteration

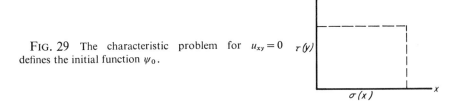

FIG. 29 The characteristic problem for $u_{xy} = 0$ defines the initial function ψ_0.

which is undertaken below. In the equivalent integral equation for the corresponding nonhomogeneous equation, it will be known as "the boundary term," referring to its role in reflecting the value of the boundary data σ and τ rather than the value of the forcing function f, or it is known as the "free term," indicating that it is a term free of integration.

We now define sets G, R, and T used in the statement of the theorem on existence and uniqueness of the solution of the characteristic boundary-value problem. Let E^1 be the set of real numbers and let

$$E^n = \{(x_1, ..., x_n) \mid x_i \in E^1, \quad i = 1, ..., n\}.$$

Let $(\alpha, \beta, \gamma, \delta) \in E^4$ be such that $\alpha < \beta$ and $\gamma < \delta$, and then let

$$G = \{(x, y, z, p, q) \in E^5 \mid (\alpha < x < \beta), \quad (\gamma < y < \delta), \quad (-\infty < z, p, q < \infty)\}$$

$$R = \{(x, y) \in E^2 \mid (\alpha < x < \beta), \quad (x < y < \delta)\}$$

$$T = \{(x, y) \in E^2 \mid (\bar{\alpha} \leqslant x \leqslant \bar{\beta}), \quad (\bar{\gamma} \leqslant y \leqslant \bar{\delta})\}$$

where $\alpha < \bar{\alpha}$, $\bar{\beta} < \beta$ and $\gamma < \bar{\gamma}$, $\bar{\delta} < \delta$. Here the closed rectangle T is contained in the open rectangle R (see Fig. 30). The region G is an infinite region in E^5 that may well be thought of as "built on an open rectangular base R." We state and prove now an existence–uniqueness theorem for the boundary-value problem (see Fig. 31)

$$\psi_{xy} = f(x, y, \psi, \psi_x, \psi_y), \qquad \psi(x, \eta) = \sigma(x), \qquad \psi(\xi, y) = \tau(y).$$

2 EXISTENCE FOR THE CHARACTERISTIC PROBLEM

Theorem[†] If

1. $(f: G \to E^1)$ and $(f \in C(G))$; $\hfill$ (6.2.1)

2. $\forall T \subset R$, $\exists M \in E^1 \ni \forall (x, y) \in T$ and $\forall (z_1, p_1, q_1), (z_2, p_2, q_2) \in E^3$

$$|f(x, y, z_2, p_2, q_2) - f(x, y, z_1, p_1, q_1)| \leqslant M(|z_2 - z_1| + |p_2 - p_1| + |q_2 - q_1|);$$
$$\hfill (6.2.2)$$

3. $\sigma: (\alpha, \beta) \to E^1$ and $\sigma \in C^1(\alpha, \beta)$ $\hfill$ (6.2.3)

$\tau: (\gamma, \delta) \to E^1$ and $\tau \in C^1(\gamma, \delta)$ $\hfill$ (6.2.4)

$\sigma(\xi) = \tau(n)$; $\hfill$ (6.2.5)

---

† The modification in Section 5 allows for a Lipschitz condition on a bounded region G.

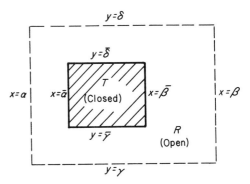

FIG. 30   The closed rectangle T is contained in the open rectangle R.

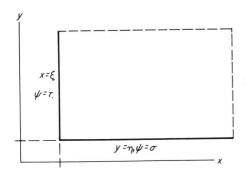

FIG. 31   The characteristic problem in characteristic coordinates.

then $\exists \psi: R \to E^1 \ni \psi_x, \psi_y, \psi_{xy} \in C(R)$ and

$$\forall x, y \in R, \qquad \psi_{xy} = f(x, y, \psi, \psi_x, \psi_y), \qquad (6.2.6)$$

$$\psi(x, n) = \sigma(x) \qquad \forall x \in (\alpha, \beta), \qquad (6.2.7)$$

$$\psi(\xi, y) = \tau(y) \qquad \forall y \in (\gamma, \delta), \qquad (6.2.8)$$

and, moreover, there is only one function $\psi: R \to E^1$ that satisfies these conditions.

Proof (a) Equivalent integral equation Integrating ψ_{xy} from (X, η) to (X, y) (see Fig. 32),

$$\psi_x(X, y) - \psi_x(X, \eta) = \int_\eta^y f(X, Y, \psi(X, Y), \psi_x(X, Y), \psi_y(X, Y)) \, dY, \quad (6.2.9)$$

103

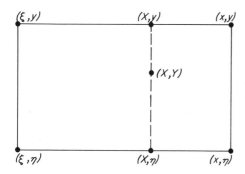

FIG. 32 Integration pattern for equivalent integral equation. Variables of integration (X, Y).

and then from (ξ, y) to (x, y),

$$\psi(x, y) = \psi(\xi, y) - \psi(x, \eta) + \psi(\xi, \eta)$$

$$+ \int_\xi^x \int_\eta^y f(X, Y, \psi(X, Y), \psi_x(X, Y), \psi_y(X, Y))\, dX\, dY$$

But, from (6.2.8), (6.2.7), (6.2.5), and (6.1.1) for any $(x, y) \in R$,

$$\psi(x, y) = \psi_0(x, y) + \int_\xi^x \int_\eta^y f(X, Y, \psi, \psi_x, \psi_y)\, dX\, dY. \qquad (6.2.10)$$

We have shown that the integral equation (6.2.10) arises from the boundary-value problem. To show that it implies the boundary-value problem, simply differentiate (6.2.10) with respect to x and then with respect to y, verifying that the differential equation is satisfied and form $\psi(\xi, y)$ and $\psi(x, n)$, verifying that the boundary conditions are satisfied.

(b) *The sequence* $\{\psi_n\}$ We define the recursion (iteration)

$$\psi^{n+1}(x, y) = \psi_0(x, y) + \int_\xi^x \int_\eta^y f(X, Y, \psi^n, \psi_x^n, \psi_y^n)\, dX\, dY \qquad (6.2.11)$$

with $\psi^0 = \psi_0(x, y)$, the "free" or "boundary term" of (6.2.10). Notice that

$$g \in C^1(R) \Rightarrow f(x, y, g, g_x, g_y) \quad \text{defines a function} \quad G \in C(R)$$

$$\Rightarrow \int_\xi^x \int_n^y G(X, Y)\, dX\, dY \quad \text{defines a function} \quad \in C^1(R).$$

Thus we have

$$\psi^n \in C^1(R) \Rightarrow \psi^{n+1} \in C^1(R).$$

Also, we have from (6.2.3), (6.2.4), and (6.1.1) that

$$\psi^0 \in C^1(R),$$

so a simple application of induction gives us that a sequence $\{\psi^n\}$ is inductively defined by (6.2.11).

(c) *Convergence* We choose any closed rectangle $T \subset R$ as shown in Fig. 33 and show that $\{\psi_n\}$ converges uniformly on T. Consider $h: T \to E^1$ defined by

$$h(x,y) = f(x,y,\psi^0,\psi_x{}^0,\psi_y{}^0)$$

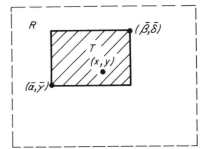

FIG. 33   Let T be any closed rectangle in the open rectangle R and let $(x,y) \in T$.

and put

$$A = \max_{(x,y)\in T} |h(x,y)|. \tag{6.2.12}$$

Then, for every $(x,y) \in T$

$$|\psi^1(x,y) - \psi^0(x,y)| \leqslant \left| \int_\xi^x \int_\eta^y f(\psi^0, \ldots) \right| dX\,dY| \leqslant A\,|x-\xi|\,|y-\eta|$$

$$|\psi_x{}^1(x,y) - \psi_x{}^0(x,y)| \leqslant A\,|y-\eta| \leqslant A(|x-\xi| + |y-\eta|) \tag{6.2.13}$$

$$|\psi_y{}^1(x,y) - \psi_y{}^0(x,y)| \leqslant A\,|x-\xi| \leqslant A(|x-\xi| + |y-\eta|).$$

Letting

$$K = \max(1, \tfrac{1}{2}|\bar{\beta}-\bar{\alpha}|, \tfrac{1}{2}|\bar{\delta}-\bar{\gamma}|) \geqslant 1, \tag{6.2.14a}$$

then

$$\frac{1}{|x-\xi|} + \frac{1}{|y-\eta|} \geqslant \frac{1}{2K} + \frac{1}{2K} = \frac{1}{K} \tag{6.2.14b}$$

and

$$|x-\xi|\,|y-\eta| \leqslant K(|x-\xi| + |y-\eta|). \tag{6.2.14c}$$

Therefore, the three inequalities (6.2.13) can be written in a "symmetric" and more convenient form—in fact, so as to form essentially a single inequality,

$$\left.\begin{array}{l} |\psi^1 = \psi^0| \\ |\psi_x{}^1 - \psi_x{}^0| \\ |\psi_y{}^1 - \psi_y{}^0| \end{array}\right\} \leqslant AK(|x-\xi| + |y-\eta|). \tag{6.2.15a}$$

In order to achieve a similarly simplified form for inequalities involving $|\psi^n - \psi^{n-1}|$, $|\psi_x{}^n - \psi_x{}^{n-1}|$, and, $|\psi_y{}^n - \psi_y{}^{n-1}|$, $n > 1$, we will need to make use of the following simple algebraic fact:

Lemma $\mu \geqslant 0, \lambda \geqslant 0 \Rightarrow (\mu+\lambda)^{n+1} - \mu^{n+1} - \lambda^{n+1} \leqslant (n+1)\mu\lambda(\mu+\lambda)^{n-1}$.

EXERCISE 1 Prove the above lemma.

Hint Expand $(\mu+\lambda)^{n+1}$ and show by first proving the inequality

$$j + 1 \leqslant n \Rightarrow \frac{n}{(n-j)(j+1)} \leqslant 1$$

that the cross terms are less than $(n+1)\mu\lambda(\mu+\lambda)^{n-1}$.

We now show by induction, first on ψ, then on ψ_x and ψ_y, that

$$\left.\begin{array}{l} |\psi^n - \psi^{n-1}| \\ |\psi_x{}^n - \psi_x{}^{n-1}| \\ |\psi_y{}^n - \psi_y{}^{n-1}| \end{array}\right\} \leqslant \frac{A}{3M}\frac{(3KM)^n}{n!}(|x-\xi| + |y-\eta|)^n, \tag{6.2.15b}$$

which, of course, is satisfied by (6.2.15a) for $n = 0$. We have

$$\psi^n - \psi^{n-1} = \int_\xi^x \int_\eta^y f(\psi^{n-1}, \ldots) - f(\psi^{n-2}, \ldots)\, dX\, dY. \tag{6.2.15c}$$

Using first the triangle inequality, then the Lipschitz condition (hypothesis 2), then the induction assumption [(6.2.15b) with n replaced by $n-1$], and then evaluating iterated integrals of polynomials, we have the following sequence of steps:

$$|\psi^n - \psi^{n-1}| \leqslant \left| \int_\xi^x \int_\eta^y f(\psi^{n-1}, \ldots) - f(\psi^{n-2}, \ldots)\, dX\, dY \right|$$

$$\leqslant \left| \int_\xi^x \int_\eta^y M\{|\psi^{n-1} - \psi^{n-2}| + |\psi_x^{n-1} - \psi_x^{n-2}| \right.$$

$$+ |\psi_y^{n-1} - \psi_y^{n-2}|\}\, dX\, dY \Big|$$

$$\leqslant \frac{A}{3M} \frac{K^{n-1}(3M)^n}{(n-1)!} \left| \int_\xi^x \int_\eta^y (|X - \xi| + |Y - \eta|)^{n-1}\, dX\, dY \right|$$

$$= \frac{A}{3M} \frac{K^{n-1}(3M)^n}{n!} \frac{1}{n+1} (|x - \xi| + |y - \eta|)^{n+1}$$

$$- |x - \xi|^{n+1} - |y - \eta|^{n+1}). \tag{6.2.16}$$

Putting $|x - \xi| = \lambda$, $|y - \eta| = \mu$ and using the lemma

$$|\psi^n - \psi^{n-1}| \leqslant \frac{A}{3M} \frac{K^{n-1}(3M)^n}{n!} [|x - \xi| |y - \eta| (|x - \xi| + |y - \eta|)^n].$$

Then, recalling (6.2.14) (where $K \geqslant 1$),

$$|\psi^n - \psi^{n-1}| \leqslant \frac{A}{3M} \frac{(3KM)^n}{n!} (|x - \xi| + |y - \eta|)^n.$$

This completes our induction for the sequence $\{\psi^n\}$.

It is not necessary to repeat the above for induction on sequences $\{\psi_x^n\}$ and $\{\psi_y^n\}$. In (6.2.16), when we differentiate with respect to x, all we do is to remove the first integration. The rest of the calculation of an upper bound for $|\psi_x^n - \psi_x^{n-1}|$ is precisely the same as that for $|\psi^n - \psi^{n-1}|$ except that some positive terms must be added after the integration of $(|X - \xi| + |Y - \eta|)^{n-1}$ in order to obtain the same upper bound. The situation is identical in finding this same upper bound is valid for $|\psi_y^n - \psi_y^{n-1}|$.

The bound now proved for $|\psi^n - \psi^{n-1}|$, $|\psi_x^n - \psi_x^{n-1}|$, $|\psi_y^n - \psi_y^{n-1}|$ is the nth term of the exponential series

$$\frac{A}{3M} e^{3KM(|x - \xi| + |y - \eta|)} \leqslant \frac{A}{3M} e^{3KM((\bar{\beta} - \bar{a}) + (\bar{\delta} - \bar{\gamma}))}. \tag{6.2.17}$$

But

$$\{\psi^n\} = \psi^0 + \sum_{i=1}^\infty (\psi^i - \psi^{i-1}) \tag{6.2.18}$$

is thus majorized by a convergent series of positive constants and $\{\psi^n\}$ is thus shown to be uniformly convergent on T. Similarly, for $\{\psi_x^n\}$ and $\{\psi_y^n\}$, we have that $\{\psi^n\}, \{\psi_x^n\}, \{\psi_y^n\}$ converge uniformly on any finite subrectangle of R. Therefore, there exist $\psi, F, G \in C(R)$ such that

$$\{\psi^n\} \to \psi, \qquad \{\psi_x^n\} \to F, \qquad \{\psi_y^n\} \to G. \qquad (6.2.19)$$

(*d*) *Limits of derivative sequences* It must now be verified that

$$F = \psi_x \qquad \text{and} \qquad G = \psi_y. \qquad (6.2.20)$$

All integrals in this chapter are intended simply to be Riemann integrals, and these integrals are interchangeable with limits of uniformly convergent sequences. The demonstration is then immediate because for every pair of points (x, y), (ξ, y) in R, and for every integer n, we have

$$\psi^n(x, y) - \psi^n(\xi, y) = \int_\xi^x (\partial\psi^n/\partial x)(t, y)\, dt,$$

so that from (6.2.19)

$$\psi(x, y) - \psi(\xi, y) = \lim_{n \to \infty} (\psi^n(x, y) - \psi^n(\xi, y)) = \lim_{n \to \infty} \int_\xi^x (\partial\psi^n/\partial x)(t, y)\, dt$$

$$= \int_\xi^x \lim_{n \to \infty} (\partial\psi^n/\partial x)(t, y)\, dt = \int_\xi^x F(t, y)\, dt,$$

or, differentiating with respect to x, $\partial\psi/\partial x = F$. Similarly, we have as promised, $\partial\psi/\partial y = G$.

(*e*) *Limit functions satisfy the integral equation* From the Lipschitz condition on f (hypothesis 2), the uniform convergence of sequences $\{\psi^n\}, \{\psi_x^n\}, \{\psi_y^n\}$ implies uniform convergence of the sequence $\{g^n\}$ where

$$g^n(X, Y) = f(X, Y, \psi^n, \psi_x^n, \psi_y^n).$$

Then from (6.2.11), the fundamental recursion, we may write

$$\psi = \lim_{n \to \infty} \psi^n = \psi_0 + \lim_{n \to \infty} \int_\eta^y \int_\xi^x f(X, Y, \psi^{n-1}, \psi_x^{n-1}, \psi_y^{n-1})\, dX\, dY$$

$$= \psi_0 + \int_\eta^y \int_\xi^x \lim_{n \to \infty} f(X, Y, \psi^{n-1}, \psi_x^{n-1}, \psi_{y-}^{n-1})\, dX\, dY$$

$$= \psi_0 + \int_\eta^y \int_\xi^x f(X, Y, \psi, \psi_x, \psi_y)\, dX\, dY$$

from the continuity of f and the considerations in (d).

(*f*) *Uniqueness* Suppose ψ_1 and ψ_2 are solutions. The student may note that for a nonlinear problem such as we have it is not always true that $\psi_2 - \psi_1$ is a solution.[†] Some technique for proving uniqueness other than that used repeatedly on linear problems in Part I must be found. Here we let W represent a bound of a particular norm of the difference of two solutions on T

$$\|\psi_2 - \psi_1\| = \max_{x,y \in T} [|\psi_2 - \psi_1| + |\psi_{2x} - \psi_{1x}| + |\psi_{2y} - \psi_{1y}|] \leqslant W \quad (6.2.21)$$

and then show that this bound can be successively improved, thus obtaining a sequence of bounds for this norm of the difference of any two solutions which converges to zero. From the integral equation (6.2.10) which is satisfied by both ψ_1 and ψ_2,

$$\psi_2 - \psi_1 = \int_\eta^y \int_\xi^x [f(X,Y,\psi_2,\psi_{2x},\psi_{2y}) - f(X,Y,\psi_1,\psi_{1x},\psi_{1y})] \, dX \, dY$$

$$\psi_{2x} - \psi_{1x} = \int_\eta^y [f(X,Y,\psi_2,\psi_{2x},\psi_{2y}) - f(X,Y,\psi_1,\psi_{1x},\psi_{1y})] \, dY$$

$$\psi_{2y} - \psi_{1y} = \int_\xi^x [f(X,Y,\psi_2,\psi_{2x},\psi_{2y}) - f(X,Y,\psi_1,\psi_{1x},\psi_{1y})] \, dX.$$

Using the Lipschitz condition (hypothesis 2) in the same manner as above,

$$|\psi_2 - \psi_1| \leqslant MW|X - \xi| \, |y - \eta|$$

$$|\psi_{2x} - \psi_{1x}| \leqslant MW|y - \eta|$$

$$|\psi_{2y} - \psi_{1y}| \leqslant MW|x - \xi|.$$

As above, we use (6.2.14c) to obtain a single bound for all three quantities,

$$\left.\begin{array}{c} |\psi_2 - \psi_1| \\ |\psi_{2x} - \psi_{1x}| \\ |\psi_{2y} - \psi_{1y}| \end{array}\right\} \leqslant MWK(|X - \xi| + |y - \eta|),$$

and note that it can now be shown by induction, using the Lipschitz condition (hypothesis 2) and the lemma proved in (c) above, that

$$\left.\begin{array}{c} |\psi_2 - \psi_1| \\ |\psi_{2x} - \psi_{1x}| \\ |\psi_{2y} - \psi_{1y}| \end{array}\right\} \leqslant \frac{W}{3} \frac{(3KM)^n}{n!} (|x - \xi| + |y - \eta|)^n. \quad (6.2.22)$$

[†] Even in case f is linear, the difference of two solutions satisfies the homogeneous equation, not the full equation.

EXERCISE 2 Prove (6.2.22) by induction.

But the expression on the right of (6.2.22), of course, is the general term of the exponential function

$$(W/3) \exp[(3KM)(|x-\xi| + |y-\eta|)],$$

and, as such, it goes to zero as n tend to infinity. We have

$$\|\psi_2 - \psi_1\| = \lim_{n \to \infty} \|\psi_2 - \psi_1\| \leqslant \lim_{n \to \infty} \frac{W}{3} \frac{(3KM)^n}{n!} (|x-\xi| + |y-\eta|)^n = 0,$$

so we have $\psi_2 - \psi_1 = 0$. This completes the proof of our theorem.

3 COMMENTS ON CONTINUOUS DEPENDENCE AND ERROR BOUNDS

In the above ψ_2 and ψ_1 may be thought of as corresponding to different initial functions ψ_0 [see (6.2.10)] and, therefore, to different boundary-value functions $\sigma(x), \tau(y)$. A similar argument then shows the solution depends continuously in maximum norm on ψ_0.

Error bounds can be conveniently derived by computing a bound for $|\psi^n - \psi^{n+m}|$ and letting $m \to \infty$ so as to give a bound for $|\psi^n - \psi|$. We do not go through this because we have intended only to give an existence and uniqueness theorem, not a computational procedure. Such function iterations (as opposed to numerical or finite-dimensional vector iterations) have so far appeared not to be generally useful as computational procedures, though we believe the last of this story is yet to be written. Some serious concentration, however, will be necessary on the properties of "computability" before such methods can become practical computation procedures. Of course, it is immediately evident that the exponential series offers a convenient remainder term, so no special considerations need be given beyond that of (6.2.15b).

4 AN EXAMPLE WHERE THE THEOREM AS STATED DOES NOT APPLY

The following is an example of a simple characteristic boundary-value problem which has a unique solution and to which our theorem as so far stated is not applicable. It shows clearly the weakness in asking that a function satisfy a Lipschitz condition in an unbounded region like G and thus shows the lack of strength of this theorem over the one ordinarily quoted for ordinary differential equations.

Let $R = [a, b] \times [c, d]$ and let $g: [a, b] \to E^1$, $\tau: [c, d] \to E^1$ such that $g \in C^1 [a, b]$, $g(a) = \tau(c)$, and for all $(x, y) \in R$

$$g'(x)(y - c) \neq 1. \tag{6.4.1}$$

We seek solution for the following problem,

$$
\begin{aligned}
\forall (x, y) \in R, \qquad & \psi_{xy}(x, y) = \psi_x{}^2(x, y) \\
\forall x \in a, b, \qquad & \psi(x, c) = g(x) \\
\forall y \in c, d, \qquad & \psi(a, y) = \tau(y).
\end{aligned} \tag{6.4.2}
$$

Since the region G is not bounded, even this simple function,

$$f(x, y, z, p, q) = p^2,$$

does not satisfy a Lipschitz condition there. To see this, let M be an arbitrary positive real number and note that whatever value M may take, we can find

$$(z_1, p_1, q_1), \quad (z_2, p_2, q_2) \in E^3$$

such that

$$z_1 = z_2, \qquad q_1 = q_2 \quad \text{and} \quad 2p_2 > 2p_1 > M$$

for every $(x, y) \in R$. By the mean-value theorem we have for some ξ such that $p_1 < \xi < p_2$,

$$
\begin{aligned}
|f(x, y, z_2, p_2, q_2) &- f(x, y, z_1, p_1, q_1)| \\
&= |p_2{}^2 - p_1{}^2| \\
&= 2\xi |p_2 - p_1| > 2p_1 |p_2 - p_1| > M |p_2 - p_1| \\
&= M\{|z_2 - z_1| + |p_2 - p_1| + |q_2 - q_1|\}.
\end{aligned}
$$

However, a solution of the boundary-value problem is

$$\psi(x, y) = \tau(y) + \int_a^x \frac{g'(s)\, ds}{1 - g'(s)(y - c)}, \tag{6.4.3}$$

as one may directly verify. It is arrived at by the following argument which also shows uniqueness:

For fixed x, $a \leqslant x \leqslant b$, define $h(y) = \psi_x(x, y)$ for all $y \in [c, d]$. On $[c, d]$, the function h satisfies the initial-value problem

$$h'(y) = h^2(y), \qquad h(c) = g'(x). \tag{6.4.4}$$

That this problem has a unique solution follows from the fact that if $H(y) = h_1(y) - h_2(y)$ is the difference of any two solutions, it satisfies the linear differential equation initial-value problem

$$H'(y) = [h_1(y) + h_2(y)] H(y), \qquad H(c) = 0,$$

whose unique solution is given by $H(y) = 0$. The unique solution of (6.4.4) is

$$h(y) = \frac{g'(x)}{1 - g'(x)(y - c)}$$

for all $y \in [c, d]$. Therefore, since x was arbitrary,

$$\psi_x(x, y) = \frac{g'(x)}{1 - g'(x)(y - c)}$$

for all $(x, y) \in R$. Now for fixed $y, c \leqslant y \leqslant d$, define $e(x) = \psi(x, y)$ for $a \leqslant x \leqslant b$. From the above, it is seen that e satisfies the initial-value problem,

$$e'(x) = \frac{g'(x)}{1 - g'(x)(y - c)}, \qquad e(a) = \tau(y),$$

whose unique solution is given by

$$e(x) = \int_a^x \frac{g'(s)}{1 - g'(s)(y - c)} \, ds + \tau(y).$$

Thus, since y was arbitrary, $\psi(x, y)$ as given by (6.4.3) is the only solution of our boundary-value problem (6.4.2).

It is evident that the difficulty lies [see (6.2.2)] in requiring that the Lipschitz condition be satisfied in an infinite region.

5 A THEOREM USING THE LIPSCHITZ CONDITION ON A BOUNDED REGION IN E^5

We now redefine G and T (see Section 1 on notation) as follows:

$$G = \{(x, y, z, p, q) \in E^5 \,|\, (\alpha < x < \beta), \quad (\gamma < y < \delta),$$

$$(|z - z_0| < Z), \quad (|p - p_0| < P), \quad (|q - q_0| < Q)\}, \qquad (6.5.1)$$

$$T = \{(x, y, z, p, q) \in E^5 \,|\, (\bar{\alpha} \leqslant x \leqslant \bar{\beta}), \quad (\bar{\gamma} \leqslant y \leqslant \bar{\delta}),$$

$$(|z - z_0| \leqslant \bar{Z}), \quad (|p - p_0| < \bar{P}), \quad (|q - q_0| \leqslant \bar{Q})\} \qquad (6.5.2)$$

where

$$\alpha < \bar{\alpha}, \qquad \bar{\beta} < \beta, \qquad \gamma < \bar{\gamma}, \qquad \bar{\delta} < \delta \qquad \bar{Z} < Z, \qquad \bar{P} < P, \qquad \bar{Q} < Q.$$

Letting

$$z_0 = \psi^0 = \psi_0(x, y) = \sigma(x) + \tau(y) - \sigma(0), \qquad p_0 = \sigma'(x), \qquad q_0 = \tau'(y),$$

we see that trivially

$$(x, y, \psi^0, \psi_x{}^0, \psi_y{}^0) \in T.$$

Of course, we still require that $f \in C(G)$, and since $T \subset G$ and T is closed, there is a positive real number A such that on T

$$|f(x, y, z, p, q)| \leqslant A. \qquad (6.5.3)$$

From (6.2.11) it follows that (6.2.13) is satisfied. That is, we have

$$|\psi^1 - \psi^0| \leqslant A|y - \eta||x - \xi|$$

$$|\psi_x{}^1 - \psi_x{}^0| \leqslant A|y - \eta| \qquad (6.5.4)$$

$$|\psi_y{}^1 - \psi_y{}^0| \leqslant A|x - \xi|.$$

But now, if we let

$$L = \min((\bar{\beta} - \bar{\alpha}), (\bar{\delta} - \bar{\gamma}), Z^{1/2}/A^{1/2}, P/A, Q/A), \qquad (6.5.5)$$

then for every (x, y) such that

$$|x - \xi| < L \qquad \text{and} \qquad |y - \eta| < L \qquad (6.5.6)$$

we have

$$|\psi^1 - \psi^0| \leqslant Z, \qquad |\psi_x{}^1 - \psi_x{}^0| \leqslant P, \qquad |\psi_y{}^1 - \psi_y{}^0| \leqslant Q \qquad (6.5.7)$$

so that $(x, y, \psi^1, \psi_x{}^1, \psi_y{}^1) \in T$ for every (x, y) in the appropriate range (6.5.6). Of course, once it is realized that

$$(x, y, \psi^k, \psi_x{}^k, \psi_y{}^k) \in T \qquad (6.5.8)$$

the same proof suffices to show that

$$(x, y, \psi^{k+1}, \psi_x^{k+1}, \psi_y^{k+1}) \in T,$$

and (6.5.8) is inductively established as a fact.

This is all that is required to improve our theorem and remove the objections which the example in Section 4 raised. We have now established the following theorem. It will be well to notice that *for a linear equation*, the Lipschitz condition being satisfied on any closed "rectangle" in E^5 however large, a unique solution can be seen to exist on any rectangle however large in E^2.

113

Theorem With G and T defined by (6.5.1) and (6.5.2), if

1. $(f: G \to E^5)$ and $f \in C(G)$;

2. $\forall T \subset G, \exists M \in E^1 \ni \forall (x, y, z_1, p_1, q_1), (x, y, z_2, p_2, q_2) \in T$

$|f(x, y, z_2, p_2, q_2) - f(x, y, z_1, p_1, q_1)| \leqslant M(|z_2 - z_1| + |p_2 - p_1| + |q_2 - q_1|)$;

3. $\sigma: (\alpha, \beta) \to E^1$ and $\sigma \in C^1(\alpha, \beta)$,

   $\tau: (\gamma, \delta) \to E^1$ and $\tau \in C^1(\gamma, \delta)$, and

   $\sigma(\xi) = \tau(\eta)$;

then, where L is defined by (6.5.5) and

$$R = \left\{ (x, y) \,\middle|\, |x - \xi| < L, \quad |y - \eta| < L \right\},$$

$\exists \psi: R \to E^1 \ni \psi_x, \psi_y, \psi_{xy} \in C(R)$, and, in R, (6.2.6), (6.2.7), and (6.2.8) are satisfied.

So far as we can determine, the theorem stated in Section 2 is the only one given in the classical literature. To improve the theorem greatly we had only to see to it that the integral equation (6.2.10) mapped a given bounded region G back into itself. In terms of the theory of contractive maps this would have been obvious, but the classicists did not think in such terms. The "finiteness" involved in the theorem as now modified, both as to the assumption of the Lipschitz condition on a bounded region and to the existence of a solution on a specified rectangle satisfying specified bounds, has a flavor that should be contrasted with the "looser" classical tastes. It may be noted that in the newer version of the theorem a maximum principle is expressed for the solution (and its first derivatives) which says that the maximum of the solution cannot differ by more than the amount Z from the maximum on the boundary (and its first derivatives by more than P and Q).

6 EXISTENCE THEOREM FOR THE CAUCHY PROBLEM OF THE NONHOMOGENEOUS (NONLINEAR) WAVE EQUATION IN E^2

We return now to the context of the classical existence theorem of Section 2; i.e., we again use the definitions of G, R, and T given in Section 1 (see Fig. 30). It will be remembered that there $G \subset E^5$ is an open region built on the open rectangle $R \subset E^2$ and T is any closed subrectangle of R.

Theorem Let $q_1, q_2: [t_1, t_2] \to R^1$; $F: (\alpha, \beta) \to R^1$, and $G: (\gamma, \delta) \to R^1$. Let $q_1, q_2 \in C^1(t_1, t_2)$, where $q_1(t_1) = \alpha, q_2(t_1) = \gamma, q_1(t_2) = \beta$, and $q_2(t_2) = \delta$, and let $F \in C^1(\alpha, \beta)$, $G \in C^1(\gamma, \delta)$. Let q_1, q_2 be strictly monotone. Then let the curve segment c be represented (parametrically, as a function of x, or as a function of y) in any one of the following three forms, assumed to be equivalent (see Fig. 34):

$$c = \{(x,y) \in R^2 \,|\, x = q_1(t), \quad y = q_2(t), \quad t_1 < t < t_2\}, \qquad (6.6.1)$$

$$c = \{(x,y) \in R^2 \,|\, y = F(x)\}, \qquad (6.6.2)$$

$$c = \{(x,y) \in R^2 \,|\, x = G(y)\}. \qquad (6.6.3)$$

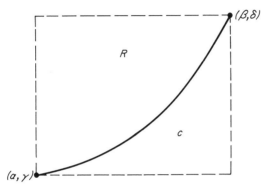

FIG. 34   The Cauchy problem is set on a curve segment c between corners (α, γ), (β, δ) of the open rectangle R.

Let R denote the smallest open rectangle bounded by $x = \alpha, \beta$ and $y = \gamma, \delta$ such that $c \subset R$. If

1. $f \in C(G)$ and $\forall T \subset R, \exists M > 0 \ni \forall (x,y) \in T$

$$|f(x,y,z_1,p_1,q_1) - f(x,y,z_2,p_2,q_2)| \leqslant M(|z_1 - z_2| + |p_1 - p_2| + |q_1 - q_2|);$$

2. $\sigma: (\alpha, \beta) \to R^1$, $\tau: (\gamma, \delta) \to R^1$, and $\sigma \in C^1(\alpha, \beta), \tau \in C^1(\gamma, \delta)$; then

$$\psi: R \to E^1 \ni \psi, \psi_x, \psi_y, \psi_{xy} \in C(R) \wedge \ni \forall (x,y) \in R,$$

$$\psi_{xy} = f(x, y, \psi, \psi_x, \psi_y),$$

and $\forall (x,y) \in c$,

$$\psi_x = \sigma'(x), \qquad \psi_y = \tau'(y), \qquad \psi = \sigma(x) + \tau(y),$$

and this function is uniquely defined on R.

115

Proof We first obtain an equivalent integral equation and then simply adapt the existence–uniqueness proof for the characteristic boundary-value problem.

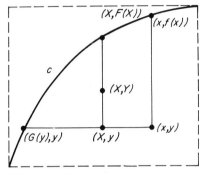

FIG. 35 Integration pattern for obtaining an integral equation equivalent to the Cauchy problem.

Noting Fig. 35, we integrate both members of the differential equation from (X, y) to $(X, F(X))$, obtaining

$$\psi_x(X, F(X)) - \psi_x(X, y) = \sigma'(x) - \psi_x(X, y)$$

$$= \int_y^{F(X)} f(X, Y, \psi, \psi_x, \psi_y) \, dY,$$

and then integrate this from $(G(y), y)$ to (x, y), obtaining

$$\sigma(x) - \sigma(G(y)) - [\psi(x, y) - \sigma(G(y)) - \tau(y)]$$

$$= \int_{G(y)}^x \int_y^{F(X)} F(X, Y, \psi, \psi_x, \psi_y) \, dX \, dY,$$

or

$$\psi(x, y) = \sigma(x) + \tau(y) - \int_{G(y)}^x \int_y^{F(X)} f(X, Y, \psi, \psi_x, \psi_y) \, dX \, dY. \quad (6.6.4)$$

It should be noted that the integral is over the region (called a "characteristic triangle") bounded by the characteristics through the point (x, y) and the data curve segment c. There are exactly four cases to consider, of which we have completed a typical one, depending on the curvature of the data segment c and the relative location of the point (x, y) at which a solution is sought (see Fig. 36), but the formula is the same for all cases. There are only four by virtue of the requirement that q_1, q_2 be strictly monotone.

With iteration set up for (6.6.4), we choose any closed rectangle $T \subset R$.

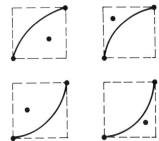

FIG. 36 The four relative positions of solution point and data segment.

Then for any $(x, y) \in T$

$$|\psi^n(x,y) - \psi^{n-1}(x,y)|$$

$$\leq \left| \int_{G(y)}^x \int_y^{F(X)} |f(X, Y, \psi^{n-1}, \psi_x^{n-1}, \psi_y^{n-1}) - f(X, Y, \psi^{n-2}, \psi_x^{n-2}, \psi_y^{n-2})| \, dX \, dY \right|$$

(6.6.5)

$$\leq \left| \int_T |f(X, Y, \psi^{n-1}, \psi_x^{n-1}, \psi_y^{n-1}) - f(X, Y, \psi^{n-2}, \psi_x^{n-2}, \psi_y^{n-2})| \, dX \, dY \right| \quad (6.6.6)$$

because the integrand and measure (or content) of the region of integration is positive or zero and the characteristic triangle is contained in the rectangle T. The remainder of the proof is identical to that given in Section 2.

EXERCISE 3 Verify that the integral equation for the Cauchy problem is the same for the four cases shown in Fig. 36.

EXERCISE 4 Modify the statement of the theorem on existence and uniqueness for the Cauchy problem so that it is required only that a Lipschitz condition be satisfied in a bounded, not an infinite, region. Use the results in Section 5 as a model.

EXERCISE 5 Avoiding the step from (6.6.5) to (6.6.6), carry out the proof of convergence of Picard iteration for (6.6.4) directly and see to what extent it is possible to sharpen the above results for the Cauchy problem.

The Riemann Method

1 THREE FORMS OF THE GENERALIZED GREEN IDENTITY

We continue here with the analysis of the nonhomogeneous wave equation (II.1). Existence and uniqueness questions, even continuous dependence, have largely been put aside, and methods of approximation are perhaps best taken up as questions of numerical analysis. The method of characteristics or a more direct finite-difference replacement technique can easily be constructed for the problems treated in Chapter 6, if the student finds the criteria available for finite-difference procedures acceptable, or he may have a try at least-squares approximation methods, as outlined for the Laplace equation in Chapter 3. Presumably, error bounds could be made available by the use of the maximum principle given in Chapter 6, Section 5. One may even wish to try, as suggested in Chapter 6, to see if the iteration method used in Chapter 6 for proving existence could be adapted to produce an effective computation procedure. It might prove particularly instructive to try to do so if finite-difference or numerical-quadrature replacements could be avoided; but, of course, these too can be invoked.

Rather than approximation methods, we turn our attention now to the representation of solutions in analytical terms. The Riemann method is a method for finding integral representations of solutions to problems like those treated in Chapter 6. These may—indeed they will—be useful in deducing important "qualitative properties" of solutions to these problems, but it should be clearly understood from the beginning that their role is representation, not numerical approximation. Obtaining numerical approximations by the route of integral representations would usually be thought to be foolish, though it must be said that some responsible workers often lay dramatic claims to success with such techniques and stoutly recommend them to us above all others. This is, however, decidedly not the intended orientation here.

We cannot succeed, of course, in obtaining integral representations for solutions of boundary-value problems for the general nonlinear equation (II.1). Instead we will be satisfied here to treat only the linear case

$$u_{xy} = -a(x,y)u_x - b(x,y)u_y - c(x,y)u + d(x,y). \qquad (7.1.1)$$

Here it will be seen that the operator L is one of those covered by the generalized Green identity (5.7.4) established in Chapter 5 provided functions a, b, and c have the required differentiability and continuity ("smoothness") properties. However, in this case, the space dimension is two only and the coefficients A^{ij} of the principle part of Lu are zero except that $A^{12} = 1$. The special circumstances permit the use of three special forms of the generalized Green identity, all of which we will find useful in this chapter.

Let the following notation then be adopted in Eq. (5.7.4):

$$x \quad \text{for} \quad x_1, \qquad y \quad \text{for} \quad x_2, \qquad P \quad \text{for} \quad P_1, \qquad Q \quad \text{for} \quad P_2, \qquad (7.1.2a)$$

$$a \quad \text{for} \quad B_1, \qquad b \quad \text{for} \quad B_2, \qquad c \quad \text{for} \quad C \qquad (7.1.2b)$$

and

$$Lu = u_{xy} + au_x + bu_y + cu. \qquad (7.1.3)$$

The adjoint of L may then be defined by

$$L^*v = v_{xy} - (av)_x - (bv)_y + cv$$

or

$$L^*v = v_{xy} - av_x - bv_y + (c - a_x - b_y)v. \qquad (7.1.4)$$

EXERCISE 1 Prove that $L^{**} = L$. Of course, this is a general property of the second-order linear operator, not just the one defined by (7.1.3).

EXERCISE 2 Determine under what special conditions the operator defined by (7.1.3) is self-adjoint (i.e., such that $L^* = L$).

The second form given for the generalized Green identity in Chapter 5 is now expressed as

$$vLu - uL^*v = P_x + Q_y \qquad (7.1.5)$$

and the three forms we will use of this identity are three forms of P and Q:

(a) $\quad P = \frac{1}{2}(vu_y - uv_y) - auv, \qquad Q = \frac{1}{2}(vu_x - uv_x) + buv,$

(b) $\quad P = \frac{1}{2}(uv)_y - u(v_y - av), \qquad Q = \frac{1}{2}(uv)_x - u(v_x - bv), \qquad (7.1.6)$

(c) $\quad P = -\frac{1}{2}(uv)_y + v(u_y + au), \qquad Q = -\frac{1}{2}(uv)_x + v(u_x + bu).$

Furthermore, let D be a region in E^2 and let A be a region with topological boundary C which is a simple closed curve and such that $A \cup C \subset D$ (see Fig. 37). Then we may write

$$\int_A (vLu - uL^*v)\, dA = \int_A (P_x + Q_y)\, dA = \int_C (Pn_x + Qn_y)\, ds$$

$$= \oint_C (P\, dy - Q\, dx)$$

where n_x, n_y are components of the outward normal $\bar{n}$ and the last integral is a line integral.

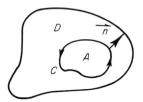

FIG. 37 C is a simple closed curve. C and its interior A are contained in D.

2 RIEMANN'S FUNCTION

The Riemann function $R_L: R^2 \times R^2 \to R^1$ is one associated with each operator of the form (7.1.3). Of course, it may be regarded as a function $R_L: R^4 \to R^1$, but this is not a fruitful viewpoint because a certain adjoint symmetry with respect to points in R^2 turns out to be of some importance here.

Definition The Riemann function $R_L = R_L((x_0, y_0); (x, y))$ is defined for all $(x_0, y_0), (x, y)$ where (x_0, y_0) and (x, y) are in an open rectangle R. It is such that for every (x_0, y_0) (regarded as fixed),

(i)
$$L(R_L(x_0, y_0; x, y)) = 0;$$
(7.2.1)

(ii)
$$\frac{\partial R_L}{\partial y}(x_0, y_0; x_0, y) + a(x_0, y) R_L(x_0, y_0; x_0, y) = 0;$$
(7.2.2)

(iii)
$$\frac{\partial R_L}{\partial y}(x_0, y_0; x, y_0) + b(x, y_0) R_L(x_0, y_0; x, y_0) = 0; \text{ and}$$
(7.2.3)

(iv)
$$R_L(x_0, y_0; x_0, y_0) = 1.$$
(7.2.4)

Condition (ii) of the definition implies that for $x = x_0$

$$R_L(x_0, y_0; x_0, y) = \exp\left[-\int_{y_0}^{y} a(x_0, s)\, ds\right]$$
(7.2.5)

and condition (iii) implies that for $y = y_0$

$$R_L(x_0, y_0; x, y_0) = \exp\left[-\int_{x_0}^{x} b(s, y_0)\, ds\right],$$
(7.2.6)

both of which equal one for $(x, y) = (x_0, y_0)$ as demanded by condition (iv). The Riemann function is thus a solution of a characteristic boundary-value problem with very specific data. The equation $Lu = 0$ being linear, it is seen by either of the theorems quoted in Chapter 6 for existence and uniqueness of the characteristic boundary-value problem that the Riemann function is categorically defined.[†]

Proposition The Riemann function is adjoint symmetric; i.e.,

$$R_L(x_0, y_0; \xi, \eta) = R_{L^*}(\xi, \eta; x_0, y_0).$$
(7.2.7)

Proof Let

$$u(x, y) = R_L(x_0, y_0; x, y)$$
(7.2.8)

and

$$v(x, y) = R_{L^*}(\xi, \eta; x, y).$$
(7.2.9)

---

† Some definitions, of course, like those of a group, ring, field, etc., are not intended to be categorical. The student of logic will recall that those that are must be justified by proving existence and uniqueness. This is the role of such proofs.

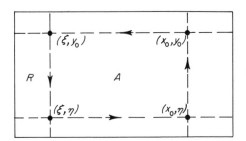

FIG. 38 Integration pattern for proof of symmetry for the Riemann function.

Then, integrating over the rectangular region A pictured in Fig. 38,

$$0 = \int_A (vLu - uL^*v)\, dA = \oint_{\delta A} (P\, dy - Q\, dx)$$

or

$$0 = \int_{(x_0,\eta)}^{(x_0,y_0)} P\, dy + \int_{(\xi,y_0)}^{(\xi,\eta)} P\, dy - \int_{(\xi,\eta)}^{(x_0,\eta)} Q\, dx - \int_{(x_0,y_0)}^{(\xi,y_0)} Q\, dx,$$

or, using form (c), then form (b), then (b) again, then (c), for components P and Q,

$$0 = \int_{(x_0,\eta)}^{(x_0,y_0)} \left[-\frac{1}{2}\frac{\partial (uv)}{\partial y} + v\left(\frac{\partial u}{\partial y} + au\right)\right] dy - \int_{(\xi,\eta)}^{(\xi,y_0)} \left[\frac{1}{2}\frac{\partial (uv)}{\partial y} - u\left(\frac{\partial v}{\partial y} - av\right)\right] dy$$

$$- \int_{(\xi,\eta)}^{(x_0,\eta)} \left[-\frac{1}{2}\frac{\partial (uv)}{\partial x} - u\left(\frac{\partial v}{\partial x} - bv\right)\right] dx$$

$$+ \int_{(\xi,y_0)}^{(x_0,y_0)} \left[-\frac{1}{2}\frac{\partial (uv)}{\partial x} + v\left(\frac{\partial u}{\partial x} + bu\right)\right] dx \qquad (7.2.10)$$

$$= -\tfrac{1}{2}uv\Big|_{(x_0,\eta)}^{(x_0,y_0)} - \tfrac{1}{2}uv\Big|_{(\xi,\eta)}^{(\xi,y_0)} - \tfrac{1}{2}uv\Big|_{(\xi,\eta)}^{(x_0,\eta)} - \tfrac{1}{2}uv\Big|_{(\xi,y_0)}^{(x_0,y_0)} \qquad (7.2.11)$$

or

$$0 = -R_L(x_0,y_0;x_0,y_0)\,R_{L*}(\xi,\eta;x_0,y_0) + R_L(x_0,y_0;\xi,\eta)\,R_{L*}(\xi,\eta;\xi,\eta)$$

$$= -R_{L*}(\xi,\eta;x_0,y_0) + R_L(x_0,y_0;\xi,\eta). \qquad (7.2.12)$$

This completes the proof. The step from (7.2.10) to (7.2.11) is by way of noting that

(1) $u = R_L$ [see (7.2.8)] and the coefficient of v in the first integral of (7.2.10) is zero according to (7.2.2), a part of the definition of R_L;

(2) $v = R_{L*}$ [see (7.2.9)] but R_{L*} is a solution of the equation $0 = L^*v = v_{xy} - av_x - bv_y + (c - a_x - b_y)v$; noting the change in sign of the coefficient of v_x over that for Lv, we see that the coefficient of u in the second integral is zero as a consequence of the definition of R_{L*};

(3) the coefficient of u in the third integral and that of v in the fourth are zero using the same reasoning and (7.2.3), a part of the definition of a Riemann function.

The technique above is far more important than the details; it will be used frequently in this chapter without repeated explanation. Without an understanding of this technique the Riemann function appears to be something of an anachronism rather than a great contribution. Of course, we will want to see this technique used for more significant purposes than simply the proving of adjoint symmetry before agreeing that the definition is really more than an anachronism. We thus proceed to complete the promised integral representation theorems. One comment first, however, may be of some interest.

Comment If L is self-adjoint, the above proposition simply expresses symmetry of the Riemann function. We have

$$R_L(x_0, y_0, ; \xi, \eta) = R_L(\xi, \eta; x_0, y_0),$$

and other propositions about the Riemann function are similarly simplified. However, in Exercise 2 of this chapter the student will have seen that the class of operators (7.1.3) which are self-adjoint is very restricted; indeed, all such operators are simply of the form

$$Lu = u_{xy} + cu, \tag{7.2.13}$$

and the study of this operator alone does not seem greatly to extend the study of the wave operator. However, self-adjoint operators are so much easier to handle than others that they often play an unexpectedly important role.

This kind of thing—the playing off of self-adjoint versus other operators—developed to be important in so many contexts in mathematics that it was eventually formulated into a very general and powerful theory. In functional analysis (a study of Banach spaces) the adjoint space A^* of a space A is the set of all continuous linear functionals on A. If A is the set $\mathscr{L}^p$ of functions whose pth power is integrable (modulo functions that agree on a set of measure zero), then $A^* \cong \mathscr{L}^q$ where $(1/p) + (1/q) = 1$, and $A^* \cong A$ if and only if $p = q = 2$. This is the special role of Hilbert spaces which may be regarded as the self-adjoint cases of Banach function spaces. The relation to the above is somewhat hidden. Every operator (linear function) on a Banach space has an adjoint, and a Green identity is often defined in these terms.

3 AN INTEGRAL REPRESENTATION OF THE SOLUTION OF THE CHARACTERISTIC BOUNDARY-VALUE PROBLEM

Theorem Let R be defined as in Chapter 6, Section 1. If

1. $a, b, c \in C^1(R)$, $d \in C(R)$, $\sigma \in C^1((\alpha, \beta))$, $\tau \in C^1(\gamma, \delta)$ and $\sigma(x_0) = \tau(y_0)$;
2. u is the unique[†] function such that $u, u_x, u_y, u_{xy} \in C(R)$ defined by

$$Lu = u_{xy} + au_x + bu_y + cu = d \tag{7.3.1}$$

and

$$u(x_0, y) = \tau(y), u(x, y_0) = \sigma(x); \quad \text{and}$$

3. R_L is the Riemann function for L in rectangle R;
then, for each $(\xi, \eta) \in R$,

$$u(\xi, \eta) = \sigma(x_0) R_L(x_0, y_0; \xi, \eta) + \int_{x_0}^{\xi} dx \int_{y_0}^{\eta} d(x, y) R_L(x, y; \xi, \eta) \, dy$$

$$+ \int_{x_0}^{\xi} (\sigma'(x) + b(x, y_0) \sigma(x)) R_L(x, y_0; \xi, \eta) \, dx$$

$$+ \int_{y_0}^{\eta} (\tau'(y) + a(x_0, y) \tau(y)) R_L(x_0, y; \xi, \eta) \, dy. \tag{7.3.2}$$

Proof Let

$$v(x, y) = R_L(x, y; \xi, \eta). \tag{7.3.3}$$

From the proposition in Section 2, we also have

$$v(x, y) = R_{L^*}(\xi, \eta; x, y), \tag{7.3.4}$$

so that

$$L^*v = 0. \tag{7.3.5}$$

Using this and $Lu = d$ [see (7.3.1)] in an integration over A as shown in Fig. 38, using forms (b) and (c) for P and Q,

---

[†] The context of the existence theorem given in Chapter 6, Section 2, with a Lipschitz condition satisfied on an unbounded region G is all that is needed here because the function f (i.e., Lu) is linear.

$$\int_A d(x,y) \, R_L(x,y;\xi,\eta) \, dx \, dy$$

$$= \int_A (vLu - uL^*v) \, dA$$

$$= \tfrac{1}{2}uv \Big|_{(x_0,y_0)}^{(\xi,y_0)} - \int_{(x_0,y_0)}^{(\xi,y_0)} v(u_x + bu) \, dx + \tfrac{1}{2}uv \Big|_{(\xi,y_0)}^{(\xi,\eta)} - \int_{(\xi,y_0)}^{(\xi,\eta)} u(v_y - av) \, dy$$

$$+ \tfrac{1}{2}uv \Big|_{(x_0,\eta)}^{(\xi,\eta)} - \int_{(x_0,\eta)}^{(\xi,\eta)} u(v_x - bv) \, dx + \tfrac{1}{2}uv \Big|_{(x_0,y_0)}^{(x_0,\eta)} - \int_{(x_0,y_0)}^{(x_0,\eta)} v(u_y + au) \, dy.$$

In the first integral on the right, $u_x + bu = \sigma'(x) + b\sigma(x)$; in the second, $v_y - av = 0$; in the third, $v_x - bv = 0$; and in the last, $u_y + au = \tau(y) + a\tau(y)$. Using (7.3.4), then, the above becomes

$$\int_{x_0}^{\xi} dx \int_{y_0}^{\eta} d(x,y) \, R_L(x,y;\xi,\eta) \, dy$$

$$= \tfrac{1}{2}uv \Big|_{(x_0,y_0)}^{(\xi,y_0)} + \tfrac{1}{2}uv \Big|_{(\xi,y_0)}^{(\xi,\eta)} + \tfrac{1}{2}uv \Big|_{(x_0,\eta)}^{(\xi,\eta)} + \tfrac{1}{2}uv \Big|_{(x_0,y_0)}^{(x_0,\eta)}$$

$$- \int_{x_0}^{\xi} R_L(x,y_0;\xi,\eta)(\sigma'(x) + b\sigma(x)) \, dx$$

$$- \int_{y_0}^{\eta} R_L(x_0,y;\xi,\eta)(\tau'(y) + b\tau(y)) \, dy$$

$$= -u(x_0,y_0)v(x_0,y_0) + u(\xi,\eta)v(\xi,\eta) - \int_{x_0}^{\xi}(\ ) \, dx - \int_{y_0}^{\eta}(\ ) \, dy$$

$$= -\sigma(x_0) R_L(x_0,y_0;\xi,\eta) + u(\xi,\eta) \cdot 1 - \int_{x_0}^{\xi}(\ ) \, dx - \int_{y_0}^{\xi}(\ ) \, dy.$$

Solving for $u(\xi,\eta)$, the representation claimed now follows.

Remark Such a theorem is called a representation theorem because it represents the solution in a certain form when it is already known that the solution exists (and is unique). Any relation or inequality derived from such a representation is, so to speak, "a posteriori" as distinguished from "a priori," in that the latter are derived in a manner that does not require the existence of a solution to the boundary-value problem. From the point of view of existence, the representation begs the question, for it represents the solution u of $Lu = d$ in

A in terms of a solution R_L of $Lu = 0$ in A for the same boundary-value problem. The redeeming feature here is that R_L is independent of the "data functions," $d(x, y)$, $\sigma(x)$, and $\tau(y)$. Thus the representation theorem can be used to assess the effect of varying these data functions or it may be used to examine qualitative properties like limits.

4 DETERMINATION OF THE RIEMANN FUNCTION FOR A CLASS OF SELF-ADJOINT CASES

In analysis, as contrasted with computations with terminating decimals, one rarely expects to be able to carry out Picard iteration in order to arrive at a solution (or an approximation of a solution) of a boundary-value problem, but it turns out to be possible here to determine the Riemann function by Picard iteration for the particular class of self-adjoint cases where $c(x, y)$ may be written as $f(x)g(y)$. Thus, for the operator

$$Lu = u_{xy} + f(x)g(y)u \qquad (7.4.1)$$

it is found that

$$R_L(x_0, y_0; x, y) = \sum_{i=0}^{\infty} \frac{(-1)^i}{i!} \left(\int_{x_0}^x f(r)\, dr \right)^i \left(\int_{y_0}^y g(r)\, dr \right)^i, \qquad (7.4.2)$$

but this result is extremely laborious to arrive at by Picard iteration, and we refrain from recording the details of this manipulation.

EXERCISE 3 Show that (7.4.2) is the Riemann function it is claimed here to be.

EXERCISE 4 Derive (7.4.2) by using separation of variables on (7.4.1) and the appropriate boundary conditions [see (7.2.5) and (7.2.6)].

If $f(x)g(y) = c$, where c is constant—i.e., if

$$Lu = u_{xy} + cu, \qquad c \in R^1 \qquad (7.4.3)$$

—then (7.4.2) becomes

$$R_L(x_0, y_0; x, y) = \sum_{i=0}^{\infty} \frac{(-1)^i}{(i!)^2} c^i (x - x_0)^i (y - y_0)^i. \qquad (7.4.4)$$

EXERCISE 5 Prove (7.4.4) directly by Picard iteration assuming (7.4.3).

126

The case (7.4.3) is known as the telegraphists' equation (see Webster [42]) and has had extremely important applications. It is not quite as special as it seems, which the following exercise will demonstrate.

EXERCISE 6 Show that where $a, b \in R^1$, the equation $u_{xy} + au_x + bu_y + cu = 0$ becomes $v_{xy} + (c - ab)v = 0$ if

$$v = u \exp[b(x - x_0) + a(y - y_0)].$$ \hfill (7.4.5)

The entire (analytic in the entire complex plane) function

$$\sum_{n=0}^{\infty} \frac{(-1)^n}{(n!)^2} \frac{z^{2n}}{2^{2n}}$$ \hfill (7.4.6)

is denoted by $J_0(z)$. It is the Bessel function [41] of order 0. Thus for L defined by (7.4.3), we may write

$$R_L(x_0, y_0; x, y) = J_0(2[c(x - x_0)(y - y_0)]^{1/2}),$$ \hfill (7.4.7)

and, for solution of the characteristic boundary-value problem,

$$u_{xy} + cu = d(x, y), \qquad c \in R^1$$ \hfill (7.4.8)

$$u(x, y_0) = \sigma(x), \qquad u(x_0, y) = \tau(y), \qquad \sigma(x_0) = \tau(y_0),$$

we may write [see (7.3.2)]

$$u(\xi, \eta) = \sigma(x_0) J_0(2[c(\xi - x_0)(\eta - y_0)]^{1/2})$$

$$+ \int_{x_0}^{\xi} dx \int_{y_0}^{\eta} d(x, y) J_0(2[c(x - \xi)(y - \eta)]^{1/2}) \, dy$$

$$+ \int_{x_0}^{\xi} \sigma'(x) J_0(2[c(\xi - x)(\eta - y_0)]^{1/2}) \, dx$$

$$+ \int_{y_0}^{\eta} \tau'(y) J_0(2[c(\xi - x_0)(\eta - y)]^{1/2}) \, dy.$$ \hfill (7.4.9)

Since (7.4.8) is a rather modest extension of the characteristic boundary-value problem for the wave equation dismissed so lightly in Chapter 2 (the differential equation is often referred to as the damped, or undamped, wave equation), (7.4.9) offers little hope for those who like to see solutions of problems written out very specifically in terms of functions that are well known to them.

5 AN INTEGRAL REPRESENTATION OF THE SOLUTION OF THE CAUCHY PROBLEM

We now return to the context of Section 3 of this chapter to present an integral representation of the solution of the Cauchy problem. The existence and uniqueness considerations for this problem were given in Chapter 6, Section 6.

Theorem Let

$$F: (\alpha, \beta) \to E^1, \qquad F \in C^1(\alpha, \beta)$$

$$G: (\gamma, \delta) \to E^1, \qquad G \in C^1(\gamma, \delta),$$

let

$$c = \{(x, y) \in E^2 \mid x \in (\alpha, \beta), \quad y = F(x)\}$$

$$= \{(x, y) \in E^2 \mid y \in (\gamma, \delta), \quad x = G(y)\},$$

and let

$$R = \{(x, y) \in E^2 \mid x \in (\alpha, \beta), \quad y \in (\gamma, \delta)\}.$$

Furthermore, let R_L be the Riemann function of

$$Lu = u_{xy} + au_x + bu_y + cu$$

where $a, b, c \in C^1(R)$ and $d \in C(R)$, and let

$$\sigma: (\alpha, \beta) \to E^1, \qquad \sigma \in C^1(\alpha, \beta)$$

$$\tau: (\gamma, \delta) \to E^1, \qquad \tau \in C^1(\gamma, \delta).$$

If u is the unique function such that $u, u_x, u_y, u_{xy} \in C(R)$ defined by $Lu = d$ in R

$$u_x = \sigma'(x), \qquad u_y = \tau'(y), \qquad u = \sigma(x) + \tau(y) \qquad \text{in} \quad c,$$

then for every (ξ, η) in R (see Fig. 39)

$$u(\xi, \eta) = \tfrac{1}{2}[\sigma(\xi) + \tau(F(\xi))] R_L(\xi, F(\xi); \xi, \eta)$$

$$+ \tfrac{1}{2}[\sigma(G(\eta)) + \tau(\eta)] R_L(G(\eta), \eta; \xi, \eta)$$

$$+ \int_{(\xi, F(\xi))}^{(G(\eta), \eta)} P \, dy - Q \, dx - \int_{G(\eta)}^{\xi} \int_{\eta}^{F(X)} d(X, Y) R_L(X, Y; \xi, \eta) \, dX \, dY$$

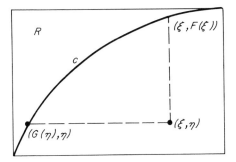

FIG. 39   Solution of the Cauchy problem at (ξ, η).

where

$$P(x, y; \xi, \eta) = \frac{1}{2} R_L(x, y; \xi, \eta) \tau'(y)$$

$$+ (\sigma(x) + \tau(y)) \left[a R_L(x, y; \xi, \eta) - \frac{1}{2} \frac{\partial R_L}{\partial y}(x, y; \xi, \eta) \right]$$

$$Q(x, y; \xi, \eta) = \frac{1}{2} R_L(x, y; \xi, \eta) \sigma'(x)$$

$$+ (\sigma(x) + \tau(y)) \left[b R_L(x, y; \xi, \eta) - \frac{1}{2} \frac{\partial R_L}{\partial x}(x, y; \xi, \eta) \right].$$

Proof We adopt the symbol $\int_\Delta$ for $\int_{G(\eta)}^{\xi} \int_{\eta}^{F(\eta)}$ and the symbol $\int$ for the corresponding line integral. The proof proceeds exactly as that of Section 3. Letting $v = R_L(x, y; \xi, \eta) = R_{L^*}(\xi, \eta; x, y)$, we have

$$\int_\Delta d(X, Y) R_L(X, Y; \xi, \eta) \, dX \, dY$$

$$= \int_\Delta (vLu - uL^*v) \, dX \, dY$$

$$= \oint (P \, dy - Q \, dx)$$

$$= \int_{(\xi, \eta)}^{(G(\eta), \eta)} Q \, dx - \int_{(\xi, F(\xi))}^{(\xi, \eta)} P \, dy + \int_{(\xi, F(\xi))}^{(G(\eta), \eta)} (P \, dy - Q \, dx)$$

$$= \tfrac{1}{2}(uv) \Big|_{(\xi, \eta)}^{(G(\eta), \eta)} - \int_{(\xi, \eta)}^{(G(\eta), \eta)} u(\partial v/\partial x - bv) \, dx$$

$$- \tfrac{1}{2}(uv) \Big|_{(\xi, F(\xi))}^{(\xi, \eta)} + \int_{(\xi, F(\xi))}^{(\xi, \eta)} u(\partial v/\partial y - av) \, dy$$

$$+ \int_{(\xi, F(\xi))}^{(G(\eta), \eta)} (P \, dy - Q \, dx).$$

Here we have simply used form (b) (of Section 1) for P and Q. It will be seen above that the coefficients of u, as they occur in two integrals, vanish by virtue of properties (7.2.2) and (7.2.3) of the Riemann function v. Then we have, using the definition of v, properties of the Riemann function, and boundary values for u,

$$
\int_\Delta d(X,Y)\,R_L(X,Y;\xi,\eta)\,dX\,dY = -u(\xi,\eta)\cdot 1
$$

$$
+ \tfrac{1}{2}[\sigma(G(\xi)) + \tau(\eta)]\,R_L(G(\eta),\eta;\xi,\eta)
$$

$$
+ \tfrac{1}{2}[\sigma(\xi) + \tau(F(\xi))]\,R_L(\xi,F(\xi);\xi,\eta)
$$

$$
+ \int_{(\xi,F(\xi))}^{(G(\eta),\eta)} (P\,dy - Q\,dx).
$$

Solving for u, the result is completed.

8

Classical Transmission Line Theory

1 THE TRANSMISSION LINE EQUATIONS

Treatments are given in elementary texts on ordinary differential equations for electrical circuits in which it is tacitly assumed that distances are negligible (or, equivalently, that propagation rates are infinite), and this allows the treatment of voltage or current, according to choice, as a function of the one variable, time, only. This will distinguish circuit theory from transmission line theory, where distance x is also an independent variable and where partial differential equations are thus involved. Only a moment's reflection will be required to convince one that distances cannot be neglected in some examples like transoceanic communications.

Consider, as in Fig. 40a, a perfect[†] resistor with resistance r and a perfect inductor with inductance l. Let $|v_2 - v_1|$ be the magnitude of voltage drop across either component when current i, depending on time t, is passed through each one. We have

$$|v_2 - v_1| = ri \qquad \text{and} \qquad |v_2 - v_1| = l\, \partial i/\partial t. \qquad (8.1.1)$$

[†] The word "perfect," of course, begs the question since a component is said to be perfect if these formulas are valid. We may say, however, for example, that $v_2 - v_1$ must vary with i only, and that it must be zero when i is zero. Then (8.1.1) must give a formula that is approximately correct for small i. To say, then, that a resistor is "perfect" is to declare this formula valid for all ranges of i it will be used for.

131

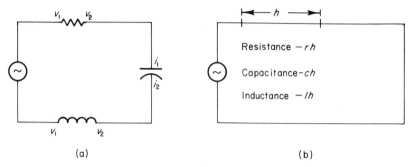

FIG. 40 (a) Circuit with resistance r, capacitance c, and inductance l represented in perfect isolated components. (b) Line with properties that give resistance r, capacitance c, and inductance l per unit length.

Likewise, for a perfect capacitor with capacitance c,

$$|i_1 - i_2| = c\, \partial v / \partial t \tag{8.1.2}$$

where v is the voltage on one side of the capacitor referenced to a fixed point on the other side.

If, as in Fig. 40b, we have a line in which each unit has resistance r, inductance l, and capacitance c, then, ignoring sign, for a length h of line,

$$v_2 - v_1 = hir + hl\, \partial i / \partial t \tag{8.1.3a}$$

and

$$i_2 - i_1 = hc\, \partial v / \partial t. \tag{8.1.3b}$$

Of course, dividing by h and letting it tend to zero, we have

$$|\partial v / \partial x| = ri + l\, \partial i / \partial t \tag{8.1.4a}$$

$$|\partial i / \partial x| = c\, \partial v / \partial t. \tag{8.1.4b}$$

Consider now a marine cable or other leaky line where current leaks out with a conductance[†] (inverse of resistance) g. It is assumed here that the voltage v is constant *across* any section of line and that the exterior sea (or other ambient conducting medium) can be treated as a shunt. Thus, taking the exterior voltage to be zero, leakage is given as gv, where it is understood, nevertheless, that the conductance g, itself, may have to be treated as a function of v. Then,

---

† Originally the leakage conductance was denoted by K (instead of g) in honor of Kirchhoff who first introduced the concept.

since the voltage is decreasing in the direction x whenever the current i is oriented in that direction, the lumped parameter description is

$$-\partial i/\partial x = gv + c\,\partial v/\partial t \qquad (8.1.5)$$

$$-\partial v/\partial x = ri + l\,\partial i/\partial t. \qquad (8.1.6)$$

These are the equations (1.7.3) which were simply quoted in Chapter 1. The result of Exercise 11 given there yields that if $i, v \in C^2(A)$, then in A

$$v_{xx} - lcv_{tt} = (rc + lg)v_t + rgv \qquad (8.1.7)$$

This, or the system (8.1.5), (8.1.6) is referred to here as the transmission line equation or system of equations. Equation (8.1.7) first appeared in a paper of Kirchhoff [25] in 1857.

2 THE KELVIN r-c LINE

For $l = 0$, Eq. (8.1.7) becomes

$$v_{xx} = (rg)v + (rc)v_t. \qquad (8.2.1)$$

Of course, this equation is of parabolic type, and transmission of information on a pure $r-c$ or $r-c-g$ line will be by diffusive spread, not by the propagation of waves as demanded by the full equation (8.1.7), which is of hyperbolic type (see Chapters 1, 2, and 4).

Of course, we can never really admit that the inductance of a line is zero because only in electrostatics, not electrodynamics, is it possible to separate completely the electric and magnetic fields. However, we will find near the end of the chapter that if the parameters r, c, l, and g are so related as to be far off a certain required balance—a balance we will later call a Heavified line—then the performance characteristics of the line may well approximate those of an $r-c$ or $r-c-g$ line.

In (8.2.1), dividing through by rc and multiplying by an integrating factor for the right member, regarded as a first-order ordinary differential operator,

$$(1/rc)(e^{(g/c)t}v)_{xx} = (1/rc)e^{(g/c)t}v_{xx} = e^{(g/c)t}((g/c)v + v_t)$$

$$= e^{(g/c)t}v_t,$$

and we have

$$DW_{xx} = W_t \qquad (8.2.2a)$$

where

$$D = 1/rc \quad \text{and} \quad W = e^{(g/c)t}v. \qquad (8.2.2b)$$

133

Thinking of our long line as doubly infinite, we specify an initial-value or Cauchy problem on the doubly infinite line $t = 0$ in the impulse form given in (4.5.2)–(4.5.4). In terms of W,

$$\text{at} \quad x = 0, \qquad \lim_{t \to 0} W = \infty \tag{8.2.3a}$$

$$\text{at} \quad x \neq 0, \qquad \lim_{t \to 0} W = 0 \tag{8.2.3b}$$

and

$$\text{for every} \quad t > 0, \qquad \int_{-\infty}^{\infty} W \, dx = C_0 \in E^1, \tag{8.2.3c}$$

the latter being a sort of generalized mass conservation condition. In terms of v [see (8.2.2b)] this condition becomes

$$\text{for every} \quad t > 0, \qquad \int_{-\infty}^{\infty} v \, dx = C_0 \, e^{-(g/c)t}. \tag{8.2.4}$$

Thus these boundary conditions on the function W provide a voltage v which initially is infinitely tall and of zero thickness, but which nevertheless has a content which leaks away exponentially with time at a rate proportional to the leakage conductance (or inversely proportional to the leakage resistance). Faraday reported (see [10, p. 508]) the insulation of Kelvin's cables to be extremely tight. Thus in (8.2.4) the exponential leakage factor could well be deleted, and Kelvin did not include this in his formulation, the concept of including a nonzero leakage in a mathematical formulation having been introduced a little later. From (4.5.6), we see that the solution of (8.2.2), (8.2.3) is

$$W = \frac{C_0}{2(\pi D t)^{\frac{1}{2}}} \exp(-x^2/4Dt) \tag{8.2.5}$$

or

$$v = \exp[-(g/c)t](rc/\pi t)^{\frac{1}{2}} C_0 \exp[-(rc/4t) x^2]. \tag{8.2.6}$$

For any given time t, the first factor on the right is a bell-shaped curve (see Fig. 41). Infinitely "sharp" at $t = 0$, it flattens with increasing t because of the factor $rc/4t$ multiplying the exponential argument $-x^2$. The next factor to the left, proportional to the square root of t, is the maximum value (value at $x = 0$) of the bell-shaped curve at any time $t \neq 0$. Thus the curve not only is being "flattened" as time passes, but it is also being "squashed down." Finally, in the case $g \neq 0$, this is all multiplied by an exponential decay factor.

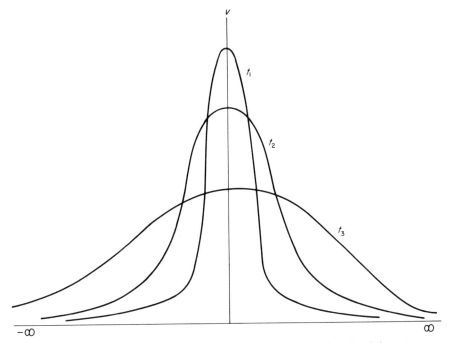

FIG. 41   An r–c impulse signal response shown in x at various fixed times t.

EXERCISE 1 From (8.2.6) prove that the Kelvin law, that the time to maximum effect is given by $t_{max} = \frac{1}{2}rcx^2$, is correct. Physiologists have long used this for small subthreshold stimuli to determine the product rc for nerve and muscle cells. In such a context it is called "electrotonic spread."

EXERCISE 2 From (8.2.6) show that the t_{max} in Exercise 1 also gives the inflection point in x (as would appear on each curve in Fig. 41) at a fixed time. Thus show that the locus in the (x, t) plane of the times to maximum effect is a parabola which is also the locus of inflection points on signals as they appear at fixed times.

Putting $l = 0$ is, at best, a risky undertaking since it eliminates one boundary condition. However, before the more general equation was known Kelvin had tested his mathematical tools on many long marine cables already in operation throughout the world. Such empirical tests, especially of the time to maximum effect, and also the theoretical derivation of it, were the burden of a paper by Kelvin in 1854.

135

3 PURE *l-c* LINE

Without taking any interest for the moment in the feasibility of constructing such circuits, we digress briefly to speak of the performance characteristics of a hypothetical *l–c* line. From (8.1.7), if $r = g = 0$, then

$$v_{xx} - (1/\theta^2)v_{tt} = 0, \tag{8.3.1a}$$

where

$$\theta = (lc)^{-\frac{1}{2}}, \tag{8.3.1b}$$

which takes us back to the opening remarks of Chapter 1; only there, of course, we had let $\theta = 1$ or, equivalently, we had let $y = \theta t$. At $t = 0$, we impose a signal

$$v(x,0) = f(x), \qquad f \in C^2(R^1), \tag{8.3.2a}$$

in a neighborhood of $x = 0$, assuming the line is previously quiet so that

$$v_t(x,0) = 0 \tag{8.3.2b}$$

at $t = 0$, and this will uniquely specify the simple solution,

$$v(x,t) = \tfrac{1}{2}[f(x+\theta t) + f(x-\theta t)]. \tag{8.3.3}$$

Let us picture (Fig. 42) our signals as functions of x at fixed times $t = 0$, $t = t_1$, $t = t_2$, etc. Of course, the peak position of the signal travels (with half its initial peak value) in both directions with constant velocity θ. In the absence of any damping, *the voltage signals at an advanced time t are exact replicas of the signals at earlier times.* We have shown that this statement is true if and only if the voltage function satisfies a wave equation, but it will be noted in Part III that these statements are equivalent only in two-dimensional space–time.

This is the prototype behavior for the phenomena of wave propagation. A similar but diffuse propagation can occur for the forced diffusion equation

$$DW_{xx} = W_t + EW_x = dW/dt, \tag{8.3.4}$$

where dW/dt is the composite derivative of W with respect to t on curves in (x, t) space with slope $dx/dt = E$. The equation (8.3.4) with impulse boundary conditions (8.2.5) yields, of course, the solution (8.2.7) with its diminishing peaks moving to right at rate E. The student will notice that the first-order operator in (8.3.4) is hyperbolic type (as first-order partial differential operators always are), and this will clarify how the effects of diffusion and propagation can occur in combination in one equation. The case (8.3.4) is so

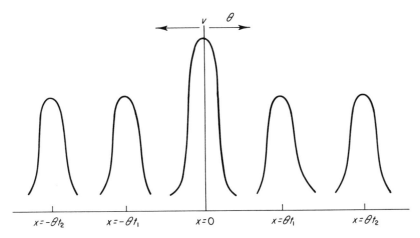

FIG. 42 Waves in a pure *l–c* line are replicas of each other. A signal impulse here is propagated with constant velocity in both directions.

different from (8.3.1a) that we never refer to it as propagation. Equation (8.3.4) is satisfied for the diffusion of a line of substance through a strip of paper when it is subjected to a constant uniform electric field, a process called electrophoresis; then E is known as the "electrophoretic mobility."

4 HEAVISIDE'S *r–c–l–g* DISTORTION-FREE BALANCED LINE

The general transmission line equation (8.1.7) which Heaviside clearly had at his disposal (and apparently thought he had introduced), since it involves the wave operator as a principal part, could clearly be expected to exhibit wave propagation properties; in fact, waves should propagate at the rate $\theta = (lc)^{-\frac{1}{2}}$ [see (8.3.1b)]. Heaviside knew this and undoubtedly appreciated more than did his contemporaries the significance of the difference between wave propagation and diffusive spread. Thus he had good reason as early as 1876 to start to think how cables could be designed so as to have large inductances. He felt strongly, as most modern applied mathematicians would, that he would fail miserably if he tried on his own to design, build, and test a high-inductance cable; so he contented himself with urging others to do so. Eventually they did, but only after long years of frustration, fighting with the head of the British telegraph system, William Preece, who wanted to increase line capacitance; but Heaviside, who lived his life by preference as a recluse, had become thoroughly hardened to going it alone. Undoubtedly, he preferred

137

not to be influenced by minds of lesser capacity, but he paid dearly for his way of life, never realizing an equitable compensation for his greatest of all communications discovery. Also, his isolation caused him to adopt a symbolism and a style of mathematical thought which was never thoroughly accepted by his contemporaries and which never influenced more than a few devoted followers. Therefore, it would be difficult now to put his discoveries in his own terms. We put them in more generally acceptable terms.

According to ideas conceived nearly a century earlier by Lagrange, whom many of the great electricians of the late nineteenth century immensely appreciated, we found in Section 4 of Chapter 7 that to study Eq. (8.1.7) one really should write it in the self-adjoint form,

$$u_{xy} + cu = 0, \qquad c \in R^1. \tag{8.4.1}$$

Here we accomplish this by a telescoping of terms like that used to obtain (8.2.4a) above for the case $l = 0$, only here we telescope terms containing factors v_t and $(v_t)_t$ rather than v and v_t. If $lc \neq 0$, we write (8.1.7) as

$$(1/lc)v_{xx} - [(v_t)_t + \tfrac{1}{2}(r/l+g/c)v_t] = (r/l)(g/c)v + \tfrac{1}{2}(r/l+g/c)v_t,$$

and multiply by the appropriate integrating factor for the expression in square brackets; we write

$$(1/lc)(e^{At}v)_{xx} - (e^{At}v_t)_t = (r/l)(g/c)ve^{At} + Ae^{At}v_t,$$

where

$$A = \tfrac{1}{2}(r/l + g/c), \tag{8.4.2}$$

and this quantity (8.4.2) will be called the exponential attenuation rate. The above can be written

$$(1/lc)(e^{At}v)_{xx} - (e^{At}v)_{tt} + Ae^{At}v_t + A^2 e^{At}v = (r/l)(g/c)e^{At}v + Ae^{At}v_t.$$

We thus have that

$$v = e^{-At}w, \tag{8.4.3}$$

where w satisfies the self-adjoint equation

$$W_{xx} - lcW_{tt} = -(lc/4)(r/l - g/c)^2 W = -H^2 W. \tag{8.4.4}$$

We call

$$H = (1/2\theta)(r/l - g/c), \tag{8.4.5}$$

where $\theta = (lc)^{-\frac{1}{2}}$, the Heavification parameter in honor of Oliver Heaviside.

138

He probably never really dealt with this parameter directly, however, because he simply did manipulations like that above (and one for the Maxwell field equations, which he admired) with the condition that

$$r/l - g/c = 0 \tag{8.4.6}$$

which is equivalent to $H = 0$ in (8.4.5). If a line has the property (8.4.6), it is called a "Heavified line," a name approved by Heaviside in a letter to his proponent Ernst J. Berg. Under these conditions we can see that (8.4.3), (8.4.4) demand that v propagate at rate $(lc)^{-\frac{1}{2}}$ without distortion, albeit at exponential attentuation rate (8.4.2). This result is quite independent of frequency so that if a line with large inductance could be made feasible, it could be used for efficient transmission of voice, which is obviously of multiple frequencies.

5 CONTRIBUTION OF DU BOIS-REYMOND AND PICARD TO THE HEAVISIDE POSITION

But even if large inductance lines could be achieved, the practical realization of Heaviside's dream demanded that if only an approximate balance (8.4.6) of resistance, inductance, leakage, and capacitance could be achieved, there would still be an approximately distortion-free transmission achieved.

The results made available to us in Sections 3 and 4 of Chapter 7 allow us to give a complete solution to (8.4.3), (8.4.4) with Cauchy initial data (8.3.2) as listed above for the pure l–c line. In fact, it allows us to solve the nonhomogeneous form of (8.4.4)

$$W_{xx} - lcW_{tt} + H^2 W = d(x, t). \tag{8.5.1}$$

Here $d(x, t)$ would be noise, say having its source from a storm at sea. Then

$$W(x,t) = \tfrac{1}{2}[f(x+\theta t) - f(x - \theta t)]$$

$$- \frac{\theta t}{2} \sum_{i=1}^{\infty} \frac{(-1)^i (H^2)^i}{i!} \int_{\frac{1}{2}(x-\theta t)}^{\frac{1}{2}(x+\theta t)} f(2r) \left(\tfrac{1}{2}(x - \theta t) - r\right)^{i-1} \left(\tfrac{1}{2}(x+\theta t) - r\right)^{i-1} dr$$

$$- \int_{\frac{1}{2}(x+\theta t)}^{\frac{1}{2}(x-\theta t)} dr \int_{\frac{1}{2}(x+\theta t)}^{\eta} d(r, s) \cdot J_0 \left[2H(\tfrac{1}{2}(x-\theta t) - r)^{\frac{1}{2}} (\tfrac{1}{2}(x+\theta t) - s)^{\frac{1}{2}}\right] ds. \tag{8.5.2}$$

Since in the summation there is no $i = 0$ term, for $d(x, t) = 0$ we conclude that

$$W = \tfrac{1}{2}[f(x+\theta t) + f(x-\theta t)] + O(H^2). \tag{8.5.3}$$

This is the result we expected, to be sure, from a glance at (8.5.1), but it is certainly reassuring to see that the distortion term due to electrical imbalance alone will be of $O(H^2)$, where H is the quantity to be designed as near to zero as possible.

From (7.4.6), (7.4.7) it is seen that

$$J_0(Y) = 1 + O(Y^2). \tag{8.5.4}$$

Thus the noise distortion term in (8.5.2) is the integral of the noise function over the forward characteristic triangle [in (x, t) space] of the given noise function $d(x, t)$ plus a function $O(H^2)$:

$$\oint d(x, t) + O(H^2). \tag{8.5.5}$$

Again the practicality of the Heaviside balanced line is seen here because for small H, any noise will simply be rather faithfully transmitted, not amplified or further distorted appreciably. The result (8.5.2) for $d(x, t) = 0$ could be obtained without using the Riemann method, either directly by Picard iteration or by separation of variables, but it would seem to be difficult to obtain (8.5.5) without its use.

6 REALIZATION

Near the turn of the century M. I. Pupin achieved large inductance lines by judiciously installing coils at discrete points on them. It may be noted that for overhead or air lines, where the leakage conductance g is altogether negligible, a virtual Heavification can be achieved if l can be made large enough since we have shown that $H \ll 1$ is all that needs to be achieved. Moreover, with large-line inductances this can be accomplished even with large resistances, resulting in an enormous saving of wire weight for overhead lines. Large coils could be placed on poles and light wires strung between them. With leakage no longer a distortion problem, lines could as well be buried as hung.

Pupin's invention had one of the most striking impacts of any invention for all of time. One reliable authority estimated the savings (in wire) involved during the first 25-year period of use in American (U.S.) experience alone to be $100,000,000. By 1936 it was estimated that 8,500,000 such loading coils were in service in the United States! Pupin was proud to say that the Germans used the phrase, *pupinizierte linien*, and the French, *les lignes pupinize*.

7 NEURONS

The axons or long tendrile portions of nerve and muscle cells behave much like marine cables, since they are buried in tissue and conducting fluid, except that because they are covered with a membrane, the leakage conductance (of ions, not free electrons) is highly nonlinear. This is the nature of the equation (1.7.3) if $g(v)$ is not taken to be a constant. Also, all the parameters depend on the radius of the axon. In the broadly applicable model of A. L. Hodgkin and A. F. Huxley, where the leakage conductance is determined from a thorough study of the empirical properties of the giant axon of the squid *Loligo*, $g(v)$ itself is formulated to satisfy an initial-value problem for a system of (essentially) ordinary differential equations which are highly nonlinear. Large inductances definitely do seem to occur on such lines and relatively large magnetic fields have been repeatedly observed propagating on nerve axons. We believe the source of such effects lies in the toroidal ion circulation patterns created through the cell membrane by the passage of a signal [30]. Signals, due to the remarkable and still mysterious properties of this thin membrane, are either passed or not passed, depending on the level of stimulus applied. Apparently, such things supply in nature, in all highly organized life, bodies that are filled with what are essentially Heavified lines. Anything less would be far too inefficient a communications system. Vertebrate nerve fibers, it turns out, might even be regarded as Pupinized lines, with inductance loads occurring at discrete points. Since myelin insulates vertebrate nerve fibers except at a discrete set of points, called nodes of Ranvier, the toroidal ion circulation patterns which apparently cause large inductance occur only in a discrete set. Some Pupin–Lagrange smearing of the inductance to give an effective continuous distribution of inductances probably takes place, but this has not been established because of the lack of an adequate nonlinear theorem to cover this phenomena. It is only in recent years that we have seriously considered the effects of inductance in electrophysiology [29].

9

The Cauchy–Kovalevski Theorem

1 PRELIMINARIES; MULTIPLE SERIES

At this point we ascertain that we have firmly in mind the precise meaning and interpretations of multiple series, especially with respect to matters relating to absolute convergence. To assist, we include here a brief resume of the material on this subject appearing in the text [1] by Apostal on pages 371–374. Proofs of theorems are not included.

Of course, the single series $\sum_{k=1}^{\infty} A_k$ is understood to be the sequence of partial sums $\{\sum_{k=1}^{n} A_k\}$, and this series, then, is said to converge if this sequence does. Double series are understood in the same way, so we must first consider the notion of a double sequence. In the following we let Ω be the set of positive integers, and we let C^1 be the set of complex numbers. Then a double sequence is a function $f: \Omega \times \Omega \to C^1$. The sequence $\{f_{pq}\}$ is said to converge to $\alpha \in C^1$ if for every $\varepsilon > 0$, there exists $N \in \Omega$ such that for every $p, q > N$,

$$|f_{pq} - \alpha| < \varepsilon.$$

We then write

$$\alpha = \lim_{p,q \to \infty} f_{pq}.$$

The double sequence $\{\sum_{j=1}^{p} \sum_{k=1}^{q} a_{j,k}\}$ of partial sums is called a double series and is denoted $\sum_{j,k=1}^{\infty} a_{j,k}$. It is said to sum to $\alpha \in C^1$ if the double sequence of partial sums converges to $\alpha \in C^1$. The double series $\sum_{j,k=1}^{\infty} a_{j,k}$ is said to converge absolutely if $\sum_{j,k=1}^{\infty} |a_{j,k}|$ converges.

Theorem If

1. $\sum_{j,k=1}^{\infty} a_{jk} = \delta$ is an absolutely convergent series,
2. $\lambda: \Omega \to \Omega \times \Omega$ is a one-to-one correspondence, and
3. $\{b_i\}$ is a sequence whose terms are defined by $b_i = a_{\lambda(i)}$, then

1. $\sum_{i=1}^{\infty} b_i$ converges absolutely to δ,
2. for every $k \in \Omega$, $\sum_{j=1}^{\infty} a_{jk}$ converges absolutely; for every $j \in \Omega$, $\sum_{k=1}^{\infty} a_{jk}$ converges absolutely, and
3. $\delta = \sum_{j=1}^{\infty} \left(\sum_{k=1}^{\infty} a_{jk} \right) = \sum_{k=1}^{\infty} \left(\sum_{j=1}^{\infty} a_{jk} \right).$

For arbitrary $N \in \Omega$, of course, an N-sequence is a function $f: \Omega^N \to C^1$ where Ω^N is the set of all N-tuples of positive integers. A sequence $\{f_{p_1, p_2, \ldots, p_N}\}$ is said to converge to $\alpha \in C^1$ if for every $\varepsilon > 0$ there exists $N \in \Omega$ such that

$$\min_{i \le N} p_i > N \Rightarrow |f_{p_1, \ldots, p_N} - \alpha| < \varepsilon.$$

If $\{a_{k_1, \ldots, k_N}\}$ is an N-sequence, then the sequence of partial sums $\{\sum_{k_1=1}^{P_1} \cdots \sum_{k_N=1}^{P_N} a_{k_1, \ldots, k_N}\}$ is called an infinite N-series and is denoted

$$\sum_{k_1, \ldots, k_N = 1}^{\infty} a_{k_1, \ldots, k_N}.$$

The N-series is said to converge to sum $\alpha \in C^1$ if the N-sequence of partial sums converges to α. The N-series is said to converge absolutely if $\sum_{k_1, \ldots, k_N}^{\infty} |a_{k_1, \ldots, k_N}|$ converges. Using induction the above theorem generalizes in the manner one would expect, allowing sums to be taken in any order for absolutely convergent multiple series.

This completes our brief résumé of the general properties of multiple series. In proving the Cauchy–Kovalevski theorem it will be well to have already in hand for quick reference some very specific information on seven multiple power series and the analytic functions they represent.

If $F: E^7 \to E^1$ is said to be an analytic function of x, y, u, p, q, r, s in a neighborhood of the point $(0, 0, 0, 0, 0, 0, 0)$, then there exists $R > 0$ such that

$$\max \{|x|, |y|, |u|, |p|, |q|, |r|, |s|\} < R \tag{9.1.1}$$

$$\Rightarrow F(x, y, u, p, q, r, s) = \sum_{\gamma_1, \ldots, \gamma_7}^{\infty} \gamma_{\gamma_1 \cdots \gamma_7} x^{\gamma_1} y^{\gamma_2} u^{\gamma_3} p^{\gamma_4} q^{\gamma_5} r^{\gamma_6} s^{\gamma_7}$$

143

converges. Let ρ be such that $0 < \rho < R$. Then the point $(\rho, \rho, \rho, \rho, \rho, \rho, \rho)$ is in the region of convergence of this series, so there exists $M > 0$ which bounds the terms of the series; i.e., there exists M such that

$$|\gamma_{\gamma_1 \cdots \gamma_7} \rho^{\gamma_1} \rho^{\gamma_2} \cdots \rho^{\gamma_7}| \leqslant M. \tag{9.1.2}$$

For any such choice of ρ and M, we have

$$|\gamma_{\gamma_1 \cdots \gamma_7}| \leqslant \frac{M}{\rho^{\gamma_1 + \cdots + \gamma_7}} \leqslant \frac{M(\gamma_1 + \cdots + \gamma_7)!}{\gamma_1! \gamma_2! \cdots \gamma_7!} \frac{1}{\rho^{\gamma_1 + \cdots + \gamma_7}} \tag{9.1.3}$$

and, defining the quantity on the right to be $\Gamma_{\gamma_1 \cdots \gamma_7}$, it follows that

$$|\gamma_{\gamma_1 \cdots \gamma_7}| \leqslant \Gamma_{\gamma_1 \cdots \gamma_7}. \tag{9.1.4}$$

This is one fact we will want to use.

Recall now that the sum of a geometric series can be written

$$a + \sum_{k=1}^{\infty} ar^k = \frac{a}{1 - r}$$

whenever $|r| < 1$. Then for $|x + y + \bar{u} + \bar{p} + \bar{q} + \bar{s} + \bar{r}| < \rho$, we have

$$\frac{M}{1 - \dfrac{x + y + \bar{u} + \bar{p} + \bar{q} + \bar{s}}{\rho}} - M = M \sum_{k=1}^{\infty} \left(\frac{x + y + \bar{u} + \bar{p} + \bar{q} + \bar{s} + \bar{r}}{\rho} \right)^k$$

$$= M \sum_{k=1}^{\infty} \left(\frac{x}{\rho} + \frac{y}{\rho} + \cdots + \frac{\bar{r}}{\rho} \right)^k$$

$$= M \sum_{\gamma_1, \ldots, \gamma_7 = 0}^{\infty} \left(\frac{x}{\rho} + \frac{y}{\rho} + \cdots + \frac{\bar{r}}{\rho} \right)^{\gamma_1 + \gamma_2 + \cdots + \gamma_7}$$

$$= \sum_{\gamma_1, \ldots, \gamma_7 = 0}^{\infty} \frac{M}{\rho^{\gamma_1 + \cdots + \gamma_7}} \frac{(\gamma_1 + \cdots + \gamma_7)!}{\gamma_7! \cdots \gamma_7!}$$

$$\times x^{\gamma_1} y^{\gamma_2} u^{\gamma_3} p^{\gamma_4} q^{\gamma_5} r^{\gamma_6} s^{\gamma_7}$$

$$= \sum_{\gamma_1, \ldots, \gamma_7 = 0}^{\infty} \Gamma_{\gamma_1 \cdots \gamma_7} x^{\gamma_1} y^{\gamma_2} u^{\gamma_3} p^{\gamma_4} q^{\gamma_5} r^{\gamma_6} s^{\gamma_7} \tag{9.1.5}$$

where $\Gamma_{\gamma_1 \cdots \gamma_7}$ is defined in (9.1.3), (9.1.4).

EXERCISE 1 Prove by induction that

$$(x+y+\cdots+\bar{r})^{\gamma_1+\gamma_2+\cdots+\gamma_7} = \sum_{\gamma_1,\ldots,\gamma_7=0}^{\infty} \frac{(\gamma_1+\cdots+\gamma_7)!}{\gamma_1!\cdots\gamma_7!} x^{\gamma_1} y^{\gamma_2} \cdots \bar{r}^{\gamma_7}.$$

This completes the preliminary material. The student will find it especially helpful now to review the materials in our "outline" (Chapters 1–5) which relate to the Cauchy problem for the Laplace equation (in both Chapters 3 and 5) although the material on the D'Alembert solution of the Cauchy problem for the wave equation in E^2 is also relevant.

2 THEOREM STATEMENT AND COMMENTS

This is one of the more difficult basic theorems in classical analysis. The setting is almost painfully classical, but the theorem provides a reference point for much modern work or at least for modern thinking. In any case, it is expected to be in the toolbag of every schoolboy.

Cauchy included a proof of this theorem in his lectures starting in 1821 in which he proceeded to put analysis on a footing more acceptable to his times. In particular he freed analysis from the logically inconsistent concept of an infinitesimal. Cauchy certainly succeeded very admirably with his task, but the impetus of his work caused the standards of clear presentation, of rigor as it were, to move upward with such grace that near the end of the century Weierstrass found the theorem of M. Cauchy we are about to present and its proof not to be entirely satisfactory. Weierstrass was concerned with concepts of analytic functions and power series; he had even sought to replace the Cauchy development of this theory for complex analytic functions by way of integral representations with one starting from the concept of power series and leading eventually to a class of theorems known as the monodromy theorems. The problem of reworking Cauchy's power series theorem for partial differential equations Weierstrass gave to his young protegé, Sonja Kovalevski.

The proof she gave has always seemed to us to be almost messy—in some ways quite difficult indeed—so every possible effort is expended here to bare the essentials of it. We seek to show existence of a unique analytic solution to a Cauchy problem given on a segment of the line $y = 0$. But this solution is really only sought in a neighborhood of the point $x = 0$ which is assumed to be on that line segment. Comment will indicate that the same proof applies equally well for any point on the segment, and, moreover, that the treatment

145

really covers any segment of any analytic curve. Further comments show that we could easily extend the treatment to higher-dimensional problems; only complications in notation and picturization would result. Power series treatments are not nearly so successful in partial differential equations as in ordinary differential equations, and the existence of solutions of initial-value problems, which are really very special for partial differential equations, is about as far as we can get in these studies. It will be noted that with Cauchy data given on $y = 0$, it is required that we can solve for u_{yy} (where u is the unknown solution sought). *This implies that $y = 0$ is not a characteristic.* However, with this noted, it becomes apparent that the results of the theorem are quite independent of type. The theorem predates any concern with type[†] or any realization of its import.

We adopt the standard notation

$$p = u_x, \qquad q = u_y, \qquad r = u_{xx}, \qquad s = u_{xy}, \qquad t = u_{yy}, \qquad (9.2.1)$$

where u is the dependent variable in a partial differential equation of second order. Where $F: E^7 \to E^1$,

$$F(x, y, u, p, q, r, s)$$

is assumed to be an analytic function at a point

$$x = 0, \qquad y = 0, \qquad u = f_0, \qquad p = p_0, \qquad q = q_0,$$

$$r = r_0, \qquad s = s_0; \qquad (9.2.2)$$

i.e., it is assumed that F can be represented as a Taylor series

$$\sum_{\gamma_1, \dots, \gamma_7}^{\infty} \gamma_{\gamma_1 \cdots \gamma_7} x^{\gamma_1} y^{\gamma_2} (u - f_0)^{\gamma_3} (p - p_0)^{\gamma_4} (q - q_0)^{\gamma_5} (r - r_0)^{\gamma_6} (s - s_0)^{\gamma_7} \quad (9.2.3)$$

which converges in some neighborhood of $(0, 0, f_0, p_0, q_0, r_0, s_0)$.

Theorem If $f_0(x), f_1(x)$ are analytic functions of x in some neighborhood of $x = 0$ and $F(x, y, u, p, q, r, s)$ is an analytic function of its arguments in some neighborhood of

$$x = 0, \qquad y = 0, \qquad u = f_0 = f_0(0), \qquad p = p_0 = f_0'(0),$$

$$r = r_0 = f_0''(0), \qquad q = q_0 = f_1(0), \qquad s = s_0 = f_1'(0),$$

[†] The limiting position of a shock, saltus discontinuity in pressure for a gas flow, is a characteristic curve. Apparently the question of type really began to take on some importance in the minds of scientists when weak shocks, which were taken for characteristics, were photographed. This occurred in rather recent times, and it explains why even Weierstrass was not concerned with type.

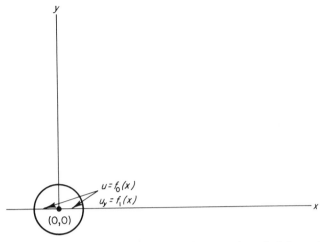

FIG. 43   The solution of the Cauchy–Kovalevski theorem is analytic in a neighborhood of the origin, but any point on the line $y = 0$ may be taken as the origin.

then the equation $t = F(x, y, u, p, q, r, s)$ possesses a unique solution $u(x, y)$ which is analytic in some neighborhood of $x = y = 0$ and takes on the initial values

$$u(x, 0) = f_0(x), \qquad u_y(x, 0) = f_1(x)$$

in that neighborhood (see Fig. 43).

Comment 1   The theorem is more general than is indicated by the restriction to $y = 0$ since an analytic transformation can be made of a region of the (x, y) plane containing the data segment $y = 0$ sending $y = 0$ into any given analytic curve. That is, if Cauchy data is prescribed on a segment of an analytic curve given by

$$\varphi(x, y) = 0,$$

then we can transform to the coordinates (see Fig. 44) given by

$$\varphi(x, y) = C_1 \in R^1, \qquad \psi(x, y) = C_2 \in R^1,$$

where ψ is any analytic function in the region such that $J(\psi, \varphi / x, y) \neq 0$. Here J is the Jacobian. Now, of course, u and $u_n = (\text{grad } u, \bar{n})$ is the given data, where $\bar{n}$ is a normal on $\varphi(x, y) = 0$.

Comment 2   As we have said, the restriction that the equation can be solved for t means $y = 0$ is not a characteristic. The fact that one can solve for t

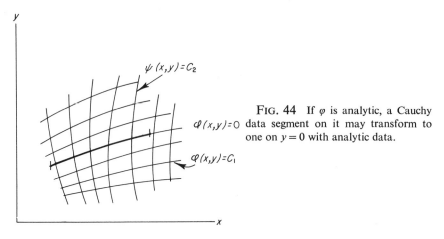

FIG. 44 If φ is analytic, a Cauchy data segment on it may transform to one on $y = 0$ with analytic data.

allows that successive coefficients in the power series solution can be computed. Convergence then is the question.

Comment 3 It will be evident from the proof that the same theorem can be proved in the same way for n independent variables $x_1, \ldots, x_n$. Instead of giving data for $y = 0$ (or $x_2 = 0$) we would give it for $x_n = 0$. Instead of looking for an analytic solution in a neighborhood of the point, $x_1 = x_2 = 0$, taken as a typical one on the data curve, we would look for a solution in a neighborhood of the point, $x_1 = x_2 = \cdots = x_n = 0$, taken as a typical one on the surface. Instead of a power series for F in seven variables, we would consider a power series in many more variables, the number depending on n. The only complications in this theorem arise in considering power series in more than one variable, the number of variables introduced beyond that cause us no additional difficulties other than bookkeeping. The initial data f_0, f_1 are now power series in more than one variable, but again this causes no difficulty. Thus we find no compelling necessity to include the additional heavy complications in notation that would be involved if we presented the theorem in n variables. With these comments our proof in two dimensions covers the n-dimensional case as well but avoids some of its complications in presentation.

3 SIMPLIFICATION AND RESTATEMENT

We transform to a new function u^* in order to restate the theorem so that, without loss of generality, the assumption on F may be simplified. We will find that F can be assumed analytic in a neighborhood of $(0,0,0,0,0,0,0)$ rather than $(0,0,f_0(0),f_0'(0),f_1'(0),f_0''(0),f_1'(0))$.

Let

$$u^*(x,y) = u(x,y) - f_0(x) - yf_1(x) - \tfrac{1}{2}y^2 f_2(x), \tag{9.3.1}$$

where

$$f_2(x) = F(x,0,f_0(x),f_0'(x),f_1(x),f_0''(x),f_1'(x))$$

EXERCISE 2 Show that at $(x,0)$ $u^* = u_y{}^* = u_x{}^* = u_{yy}^* = u_{xx}^* = u_{xy}^* = 0$.

We note that u,p,q,r,s are analytic functions of x,y,u^*,p^*,q^*,r^*,s^* about $(0,0,0,0,0,0,0)$; and t^* is an analytic function of x,y,u,p,q,r,s about $(0,0,f_0,p_0,q_0,r_0,s_0)$. Therefore, t^* is an analytic function of x,y,u^*,p^*,q^*,r^*,s^* about $(0,0,0,0,0,0,0)$. Thus we may write

$$t^* = F^*(x,y,u^*,p^*,q^*,r^*,s^*),$$

and F^* is analytic in a neighborhood of $(0,0,0,0,0,0,0)$. We will prove the Cauchy–Kovalevski theorem for this latter equation and henceforth delete all asterisk superscripts.

The Cauchy–Kovalevski theorem is now restated as follows:

Theorem If $f_0(x), f_1(x)$ are functions of x, analytic in some neighborhood of $x = 0$ and $F(x,y,u,p,q,r,s)$ is an analytic function of its arguments in some neighborhood of $x = y = u = p = q = r = s = 0$, then

$$t = F(x,y,u,p,q,r,s)$$

possesses a unique analytic solution in some neighborhood of $x = y = 0$, taking on the initial values

$$u(x,0) = f_0(x), \qquad u_y(x,0) = f_1(x)$$

in that neighborhood. Moreover, without loss of generality, it may be assumed $f_0(x) = f_1(x) = 0$ and $F(0,0,0,0,0,0,0) = 0$.

4 UNIQUENESS

For uniqueness considerations we ignore the fact that $f_0(x)$ and $f_1(x)$ may be taken to be zero. Let $u(x,y)$ be an analytic solution. Then there exists real numbers c_{ik} such that

$$u(x,y) = \sum_{i,k=0}^{\infty} c_{ik} x^i y^k \tag{9.4.1}$$

149

in some neighborhood of $(0, 0)$. Differentiation and evaluation at $(0, 0)$ gives

$$c_{ik} = \frac{u_{x^i y^k}(0, 0)}{i! \, k!}.$$
(9.4.2)

But u must satisfy the initial conditions, so $u(0, 0)$, $u_y(0, 0)$ are already given. To obtain $u_x(0, 0)$, $u_{xx}(0, 0), \ldots$, etc., we differentiate $u(x, 0) = f_0(x)$ with respect to x and evaluate at $x = 0$. All orders of derivatives of u_y with respect to x alone may also be obtained in this way from $f_1(x)$. To obtain $u_{yy}(0, 0)$ we use the differential equation. Thus

$$u_{yy}(0, 0) = F(0, 0, f_0, p_0, q_0, r_0, s_0).$$

To obtain $u_{yyy}(0, 0)$ or $u_{yyx}(0, 0)$ we differentiate F with respect to y or x, respectively, and evaluate at $x = y = 0$ using the previously obtained values for derivatives at $x = y = 0$. In this way all $u_{x^i y^k}(0, 0)$ can be developed. Thus all the coefficients c_{ik} of any analytic solution can be uniquely determined.

This completes the proof of uniqueness, which is indeed trivial for this theorem, but one should note for later use how the c_{ik}'s were formed.

5 THE FIRST MAJORANT PROBLEM

We will have established existence if we can show that the series (9.4.1) representing u converges. Let us represent the given three analytic functions as series:

$$F = \sum \gamma_{\gamma_1 \cdots \gamma_7} x^{\gamma_1} y^{\gamma_2} u^{\gamma_3} p^{\gamma_4} q^{\gamma_5} r^{\gamma_6} s^{\gamma_7},$$
(9.5.1)

$$f_0(x) = \sum a_i x^i,$$
(9.5.2)

$$f_1(x) = \sum b_i x^i.$$
(9.5.3)

If we now determine each c_{ik} in (9.4.1), representing u by the method described in Section 4, we find that c_{ik} *is a polynomial in the* a_i's, b_i's, *and* γ's, *and all coefficients in this polynomial are positive.* This important fact arises because no operations of subtractions (or divisions, for that matter) occur in the process of generating the c_{ik}'s from the a_i's and b_i's. Only additions and multiplications occur. We write

$$c_{ik} = P(a_i\text{'s}, b_i\text{'s}, \gamma\text{'s}),$$
(9.5.4)

all with positive coefficients. In [16] Garabedian presents the same proof as we are giving here but sets the problem to begin with as one for a first-order system and is then able to give an explicit formula for P showing the positive coefficients. If a student doubts this property, on which the entire proof hinges, or if he simply wishes to see P explicitly, he is invited to consult this fine text; he will find much else there that interests him.

We now replace the a_i's in f_0 (9.5.2) and the b_i's in f_1 (9.5.3) by A_i and B_i such that

$$|a_i| \leqslant A_i, \qquad |b_i| \leqslant B_i, \tag{9.5.5}$$

and we replace the γ's in F by Γ's such that

$$|\gamma_{\gamma_1 \cdots \gamma_7}| \leqslant \Gamma_{\gamma_1 \cdots \gamma_7}, \tag{9.5.6}$$

and we put

$$C_{ik} = P(A_i\text{'s}, B_i\text{'s}, \Gamma\text{'s}), \tag{9.5.7}$$

where P is the polynomial appearing in (9.5.4). As a consequence of the fact that P has positive coefficients,

$$|c_{ik}| \leqslant C_{ik}. \tag{9.5.8}$$

We now have the series (9.4.1) majorized by the series

$$\bar{u}(x, y) = \sum C_{ik} x^i y^k, \tag{9.5.9}$$

and if the latter series converges so does the former. We must be able to show that $\bar{u}$ is a solution of the altered differential equation; i.e., we must show that (9.5.9) is a solution of

$$\bar{t} = \bar{F} = \sum \Gamma_{\gamma_1 \cdots \gamma_7} x^{\gamma_1} y^{\gamma_2} \bar{u}^{\gamma_3} \bar{p}^{\gamma_4} \bar{q}^{\gamma_5} \bar{r}^{\gamma_6} \bar{s}^{\gamma_7}, \tag{9.5.10}$$

where

$$\bar{u}(x, 0) = \bar{f}_0(x) = \sum A_i x^i, \tag{9.5.11}$$

$$\bar{u}_y(x, 0) = \bar{f}_1(x) = \sum B_i x^i. \tag{9.5.12}$$

6 AN ORDINARY DIFFERENTIAL EQUATION PROBLEM

We now choose appropriate A_i's, B_i's, and Γ's. Since $f_0(x) = \sum a_i x^i$ and $f_1(x) = \sum b_i x^i$ can, without loss of generality, be chosen to be zero [see (9.1.1) and following], the A_i's and B_i's may be chosen to be any positive numbers we find convenient and the dominance (9.5.5) will still be valid. For $\Gamma_{\gamma_1 \cdots \gamma_7}$ we make the choice

$$\Gamma_{\gamma_1 \cdots \gamma_7} = \frac{M}{\rho^{\gamma_1 + \cdots + \gamma_7}} \frac{(\gamma_1 + \cdots + \gamma_7)!}{\gamma_1! \gamma_2! \cdots \gamma_7!} \tag{9.6.1}$$

where M and ρ are positive real numbers free to choice. This choice is taken because then the series (9.5.10) can be summed,

$$\bar{t} = \frac{M}{1 - \dfrac{x+y+\bar{u}+\bar{p}+\bar{q}+\bar{r}+\bar{s}}{\rho}} - M, \qquad (9.6.2)$$

as we have shown [see (9.1.5)] in the Preliminaries section. Also [see (9.1.3), (9.1.4)], the dominance (9.5.6) is thus achieved.

In its turn we replace this problem, using y/α $(0 < \alpha < 1)$ in place of y, to obtain another still more dominant series whose sum is

$$\bar{t} = \frac{M}{1 - \dfrac{x+y/\alpha+\bar{u}+\bar{p}+\bar{q}+\bar{r}+\bar{s}}{\rho}} - M \qquad (9.6.3)$$

and replace this problem by

$$\bar{t} = \frac{M}{\left[1 - \dfrac{x+y/\alpha+\bar{u}+\bar{p}+\bar{q}}{\rho_1}\right]\left[1 - \dfrac{\bar{r}+\bar{s}}{\rho_2}\right]} - M \qquad (9.6.4)$$

which dominates it for appropriate choice of M, ρ_1, and ρ_2. We now look for solutions of the form

$$u(x,y) = f(y+\alpha x) = f(z), \qquad 0 < \alpha < 1. \qquad (9.6.5)$$

The differential equation then reduces to

$$\left[1 - \frac{M}{\rho_2}(\alpha+\alpha^2)\right]f'' - \frac{\alpha+\alpha^2}{\rho_2}(f'')^2 = \frac{M}{1 - \dfrac{z/\alpha+f+(1+\alpha)f'}{\rho_1}} - M$$

$$= R(z,f,f'), \qquad (9.6.6)$$

where the function R is seen to be analytic in its arguments and with positive Taylor coefficients if it is expanded using the binomial theorem.

Constants M, ρ_1, ρ_2, have been determined in the domination of (9.6.4) over (9.6.3). However, we are still free to choose $0 < \alpha < 1$. We choose α such that

$$\alpha + \alpha^2 < \rho_2/M; \qquad (9.6.7)$$

i.e., so that

$$1 - \frac{M}{\rho_2}(\alpha+\alpha^2) > 0. \qquad (9.6.8)$$

Then the differential equation (9.6.6) becomes

$$P_1(f'')^2 - P_2 f'' + R = 0$$

where P_1, P_2, and R have positive coefficients. Then certainly

$$f'' = \frac{P_2 \pm (P_2{}^2 - 4P_1 R)^{\frac{1}{2}}}{2P_1},$$

but it is necessary only to determine one solution so we choose

$$
\begin{aligned}
f'' &= \frac{P_2 - (P_2{}^2 - 4P_1 R)^{\frac{1}{2}}}{2P_1} = \frac{P_2}{2P_1} - \frac{P_2}{2P_1}\left(1 - \frac{4P_1}{P_2{}^2}R\right)^{\frac{1}{2}} \\
&= \frac{P_2}{2P_1} - \frac{P_2}{2P_1}\left(1 - \frac{1}{2}\frac{4P_1}{P_2{}^2}R - (\text{pos. coeff.})\,R^2 \ldots\right) \\
&= S.
\end{aligned}
\tag{9.6.9}
$$

Here S is analytic with positive coefficients. As previously stated, we take $f(0) = f'(0) = 0$; this is sufficient to determine a solution of an equation of this type. Moreover, $f(z)$ is analytic since S is analytic.

We have

$$f'' = S = \sum a_{ijk} z^i f^j (f')^k,$$

$$f''(0) = a_{000}, \qquad \text{a positive constant,}$$

and, therefore,

$$f^{(i)}(0) = \text{a positive constant.} \tag{9.6.10}$$

But now we choose

$$A_i = \bar{A}_i \alpha^i, \qquad B_i = (i+1)\bar{A}_{i+1}\alpha^i$$

where

$$\bar{A}_i = f^{(i)}(0).$$

This choice is suitable since, thereby, A_i and B_i are positive and a dominant problem is solved by f. This completes the proof.

7 REMARKS AND INTERPRETATIONS

(1) We repeat for emphasis that the theorem would not be applicable if the Cauchy data were specified on a characteristic. The fact that we can solve

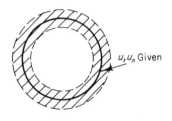

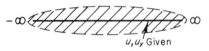

FIG. 45 The Cauchy–Kovalevski theorem does not contradict the known facts for elliptic type.

for $t = u_{yy}$ implies that $y = 0$ is not a characteristic. There is thus no contradiction here with what we have learned from the theory of characteristics in Chapter 1.

(2) In the case of elliptic equations there is no contradiction with what we know about the Dirichlet problem, where one function u, not u and u_n, are given around a closed boundary. If u_n happens to be given in such a way as to agree with the value determined by specification of u alone, then the unique analytic solution given by the Cauchy–Kovalevski theorem may be valid at every point of the interior. Otherwise, it will be determined only in a neighborhood of each point of the boundary (see Fig. 45), and this will not cover the interior. In the case where the circle is thought of as having an infinite radius, we have u and u_y given on an entire straight line. Again, unless the choice of u_y is propitious, the solution determined in the theorem will not be valid in the entire plane but only in a neighborhood of the line.

It should also be said that solutions of the Dirichlet problem are not expected to be analytic in a neighborhood of the boundary though they are in

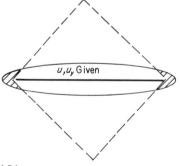

FIG. 46 Only a twice differentiable solution is determined in the characteristic triangle. An analytic solution is determined in the oblong region shown.

the interior. This will emphasize the very special role of the Cauchy–Kovalevski theorem and of the Cauchy problem for elliptic type.

(3)   In the case of hyperbolic equations, there is no contradiction between say the D'Alembert solution of the wave equation in E^2 and the power series solution shown to be uniquely determined by the Cauchy–Kovalevski theorem.

The latter solutions are analytic. The role of characteristics is to yield solutions for nonanalytic data. If analytic data u and u_y are given on a closed interval of the real axis, the neighborhood of the endpoint where the analytic solution is determined will extend outside the characteristic rectangle (see Fig. 46).

III

Some Classical Results for the Laplace and Wave Equations in Higher-Dimensional Space

10

A Sketch of Potential Theory

1 UNIQUENESS OF THE DIRICHLET PROBLEM USING THE DIVERGENCE THEOREM

Of our "outline" material (i.e., Part I) one may now want to review Chapter 3, Sections 1, 4, and 5, and Chapter 5, Section 6. The definition of what we will mean by regular solutions of the Dirichlet problem for the Laplace equation was given in Chapter 3, Section 1, and a proof of uniqueness for this problem, as well as that related to the Neumann and mixed problems stated on a rectangle in E^2, was given in Chapter 3, Section 4. This simple proof can be immediately modified to give the same results for the same problems only stated on a (hyper)rectangular prism in E^n, $n \geqslant 2$, by simply making the replacements

$$u_{xx} + u_{yy} \quad \text{by} \quad \sum_{i=1}^{n} u_{x_i x_i} \quad \text{and} \quad u_x{}^2 + u_y{}^2 \quad \text{by} \quad \sum_{i=1}^{n} u_{x_i}^2$$

and adjusting intermediate steps accordingly.

EXERCISE 1 Prove the uniqueness of regular solutions for the Dirichlet and mixed problems (also uniqueness of the Neumann problem up to a constant) on a rectangular prism using only the Fubini theorem, which allows a volume or area integral to be written as an iterated integral.

For any Green region (a region for which the divergence theorem is valid) A in E^n, the same proof of uniqueness for regular solutions can again be utilized. Now instead of simply using the Fubini theorem to pass from the volume integrals to the surface integrals and then finding that these integrals vanish, the same step is accomplished by application of the divergence theorem (5.6.3).

EXERCISE 2 Prove the uniqueness of regular solutions for the Dirichlet and mixed problems (also uniqueness of the Neumann problem up to a constant) on any Green region.

2 THE THIRD GREEN IDENTITY IN E^3

In Chapter 5, Section 6, two closely related forms were given of the divergence theorem, called the first and second Green identities by Kellogg [24]. They are stated in Eq. (5.6.6) and (5.6.7) and are valid in integral form (after application of the divergence theorem) for all functions $u, v \in C^2(A)$ and $u, v \in C^1(A \cup \delta A)$, where A is any Green region in E^n. We will develop here what Kellogg styled the third Green identity, which is not nearly so evident or so nearly a *simple* restatement of the divergence theorem.

Let $P, Q \in A$, a Green region in E^n, and consider the region[†] $A - S(\rho, P)$, (see Fig. 47) where $S(\rho, P)$ is a spherical neighborhood of radius ρ about the point P. Let r_{PQ} be the distance between points P and Q. Where

$$P = (x_1{}^0, x_2{}^0, \ldots, x_n{}^0), \qquad Q = (x_1, x_2, \ldots, x_n), \qquad (10.2.1)$$

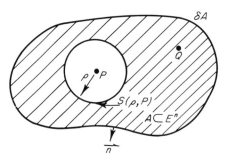

FIG. 47 A region A in E^n with boundary δA but with a spherical neighborhood $S(\rho, P)$ of radius ρ about $P \in A$ deleted. $Q \in A - S(\rho, P)$.

[†] Excluding also the boundary of $S(\rho, P)$, though this notation hides it.

we write

$$r_{PQ} = \left(\sum_{i=1}^{n} (x_i - x_i^0)^2 \right)^{1/2}. \tag{10.2.2}$$

The following then are salient facts in $A - S(\rho, P)$ about the Laplace operator,[†]

$$\Delta = \sum_{i=1}^{n} \partial^2(\ )/\partial x_i^2; \tag{10.2.3}$$

for $n = 2$,

$$\Delta(\ln r_{PQ}) = -\Delta(\ln(1/r_{PQ})) = 0, \tag{10.2.4}$$

and for $n \geqslant 3$,

$$\Delta(1/r_{PQ}^{n-2}) = 0. \tag{10.2.5}$$

The functions $\ln(1/r_{PQ})$, for $n = 2$, and $1/r_{PQ}^{n-2}$, for $n \geqslant 3$, are known as "fundamental solutions" of the Laplace equation, not only because of the fundamental role they play in deriving the important third identity but also for their role in deriving integral representations of solutions of Dirichlet and Neumann problems.

EXERCISE 3 Prove (10.2.4).

EXERCISE 4 Prove (10.2.5).

EXERCISE 5 Using a special choice of the function v in the second identity (5.6.7) show that for any function u harmonic in A

$$\int_{\delta A} \partial u/\partial n \, ds = 0. \tag{10.2.6}$$

This gives the compatibility condition required on data for the Neumann problem.

In the second identity (5.6.7) applied to $A - S(\rho, P)$ we now let $n = 3$ and take

$$v = 1/r_{PQ} \tag{10.2.7}$$

[see (10.2.2)]. Using (10.2.5), this becomes

$$-\int_{A-S} (1/r_{PQ}) \, \Delta u \, dA_Q = \int_{\delta A \cup \delta S} [u(\partial/\partial n)(1/r_{PQ}) - (1/r_{PQ})(\partial/\partial n)(u)] \, ds_Q \tag{10.2.8}$$

[†] Since derivatives are with respect to components x_i of Q, not x_i^0 of P, we here regard P as a "fixed," and Q as a "variable," point.

where, following the accepted classical practice, Q subscripts are listed on dA and dS to remind us that it is the components of Q, not P, which are considered variable here [see (10.2.1) and (10.2.3)] and over which the integration is intended. Of course, the boundary integrals in (10.2.8) can be written as the sum of integrals over the two portions, δS and δA, of the boundary. We now shrink δS to the point P and this results in the third Green identity as it appears in E^3,

$$u(P) = -(1/4\pi) \int_A (1/r_{PQ}) \Delta u \, dA_Q + (1/4\pi) \int_{\delta A} (\partial u/\partial n)(1/r_{PQ}) \, ds_Q$$

$$-(1/4\pi) \int_{\delta A} u \, \partial(1/r_{PQ})/\partial n) \, ds_Q. \tag{10.2.9}$$

There are several points to be verified in establishing (10.2.9), all of which hinge on realizing that ρ is constant on δS and that, since the outer normal to the region $A - S$ is the inner normal to S,

$$\partial(\ )/\partial n = -\partial(\ )/\rho.$$

For example, we can note that

$$\int_{\delta S} (1/r_{PQ})(\partial u(Q)/\partial n) \, ds_Q = -(1/\rho) \int_{\delta S} (\partial u(Q)/\partial \rho) \, ds_Q$$

$$= -(1/\rho)(\partial u(Q^*)/\partial \rho) \int_{\delta S} ds_Q$$

$$= 4\pi\rho(\partial u(Q^*)/\partial \rho), \tag{10.2.10}$$

where Q^* is some point belonging to δS. This is simply the mean-value theorem for integrals. But no matter what point Q^* is on δS, as ρ approaches zero, Q^* approaches P; perhaps this approach is on some strangely curled space path, but it is nonetheless valid, and, moreover, $\partial u/\partial \rho$ is assumed continuous $[u \in C^2(A)]$. We thus have that (10.2.10) goes to zero as ρ diminishes.

EXERCISE 6 Show by an argument analogous to the above that

$$\lim_{\rho \to 0} \int_{\delta S} u(\partial/\partial n)(1/r_{PQ}) \, ds_Q = 4\pi u(P).$$

Note that the 4π arises simply as the area of a unit sphere in E^3.

EXERCISE 7 Prove that

$$\lim_{\rho \to 0} \int_S ((1/r_{PQ}) \Delta u) \, dA_Q = 0.$$

Part of the importance of the third identity (10.2.9) lies in the interpretations that can be given to each of the three terms composing u. Since the function $1/r_{PQ}$ satisfies the Laplace equation, it is said to be a potential function; any linear combination or limit of linear combinations at various points Q is likewise, so $\int_A (1/r_{PQ}) \Delta u) \, dA_Q$ is a potential function at points outside A. It is said to be a potential due to a mass (or charge) distribution Δu in the volume A. Here the term "potential" is used in the very specific sense that it is a function φ such that grad $\varphi = F$ and F is the gravitational (or electrostatic) force created by the presence of mass (or charge) distribution Δu in A. Similarly, $\int_{\delta A} ((\partial u/\partial n)(1/r_{PQ})) \, ds_Q$ is a potential due to the mass (or charge) density distribution $\partial u/\partial n$ on the surface (or lamina) δA.

EXERCISE 8 Prove that the potential of a unit mass or positive charge in E^3 is given by $1/r_{PQ}$.

To understand the third member of (10.2.9) in these terms, consider the potential due to a pair of charges of strength $+e$ at $Q_1 = (a, 0, 0)$ and $-e$ at $Q_2 = (-a, 0, 0)$ (see Fig. 48). At the point $P = (x, y, z)$, we have

$$u_{ae}(P) = \frac{e}{r_{PQ_1}} - \frac{e}{r_{PQ_2}}$$

$$= e \left(\frac{1}{[(x-a)^2 + y^2 + z^2]^{1/2}} - \frac{1}{[(x+a)^2 + y^2 + z^2]^{1/2}} \right)$$

From the mean-value theorem, there is a ξ_0 between $-a$ and $+a$ such that

$$u_{ae}(P) = e \frac{\partial}{\partial \xi} \left(\frac{1}{[(x-\xi_0)^2 + y^2 + z^2]^{1/2}} \right) 2a.$$

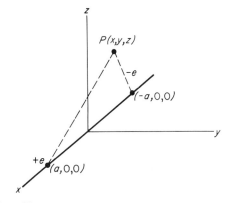

FIG. 48   The potential due to a charge dipole.

Now let $a \to 0$ and $e \to \infty$ in such a way that $\mu = 2ea$, called the dipole moment, is constant. Then the potential due to a dipole oriented in the x direction and located at the origin with moment μ is given by

$$u(P) = \lim_{\substack{a \to 0 \\ e \to \infty}} u_{ae}(P) = \mu \frac{\partial}{\partial \xi}\left(\frac{1}{r_{PQ}}\right)$$

where $Q = (0,0,0)$.

EXERCISE 9 Verify that $\Delta u(p) = 0$ at points $P \neq Q$.

It is not difficult to imagine a line or curve distribution of dipoles. In modern life the most familiar example is that of the simple television antenna, a folded loop of pipe or simply a piece of lead-in wire cut to an appropriate resonating length and spliced at the ends, but, unfortunately this is a current dipole, not a charge dipole as we now consider. The third component of u as given in (10.2.9), $\int_A (u \, \partial(1/r_{PQ})/\partial n) \, ds_Q$, is the potential due to a dipole moment distribution u on the surface (or lamina) δA.

Thus it is that the third identity declares that *any* function $u \in C^2(A)$, $C^1(A \cup \delta A)$, can be represented as the sum of a volume potential, a surface mass or charge potential, and a surface dipole potential. Each of these entities was, at least once, thought to have great physical importance,[†] and it was striking indeed that any such general function could be written as a sum of them. Before the student begins to feel that there is much more here than could reasonably be believed, let us note that a potential satisfies the Laplace equation only at those points exterior to points of mass or charge or dipole causing it. However, at such points they still satisfy the Poisson or non-homogeneous Laplace equation, and there is thus no denying that the third identity does represent a remarkable, if thoroughly elementary, result. Its limitation, we will find in the next chapter, lies in using it on dynamic problems, but even this fault can be rectified, and was by Kirchhoff.

A comment of further, though not so general, interest is found by putting $\Delta u = 0$ in (10.2.9). Then it can be seen that any harmonic function in E^3 is determined by its value and the value of its normal derivative on the boundary. From earlier uniqueness proofs, however, which were simply other uses of the divergence theorem, we know that not so much is needed; i.e., we know that u and u_n on the boundary are closely related for a harmonic function, one of them being all that can be independently specified.

---

† Perhaps they still do for statics; but statics is now a much less important subject than it was once thought to be.

3 USES OF THE THIRD IDENTITY AND ITS DERIVATION FOR E^n, $n \neq 3$

We now take the very special case of (10.2.9) that $A = S(r, P)$ and $\Delta u = 0$ in A. Then

$$u(P) = (1/4\pi r) \int_{\delta S} \partial u/\partial n \, ds + (1/4\pi r^2) \int_{\delta S} u \, ds, \qquad (10.3.1)$$

but from (10.2.6) (Exercise 5) this is

$$u(P) = (1/4\pi r^2) \int_{\delta S} u \, ds. \qquad (10.3.2)$$

This equation is known as the Gauss mean-value theorem. Since $4\pi r^2$ is the surface area of $S(r, P)$, it states that the value of a harmonic function at a point is equal to its mean-value on the surface of any sphere with that point as center. This gives one a picture, not otherwise obtainable perhaps, of the nature of a harmonic function, each point simply being the average all of those at equal distances from it—and for any distance. Along with the closely related maximum principle, it shows a property for harmonic functions which is very like one for linear functions. One finite-difference version says that the value at a point is the average of those in the four compass directions, and this has been much favored in the past to be repeated successively as a self-correcting procedure (see Chapter 3).

It will be noted that in deriving (10.3.2), all terms involving derivatives of u on the boundary dropped out. It turns out, as one would expect, that this formula can be rederived now without assuming that $u \in C^1(A \cup \delta A)$ and we can then prove that the maximum principle (3.5.6) is valid for every $u \in C^2(A)$, $u \in C(A \cup \delta A)$, $A \subset E^3$.

EXERCISE 10 Give an indirect proof that for every $u \in C^2(A)$, $C(A \cup \delta A)$,

$$\max_{A \cup \delta A} u = \max_{\delta A} u$$

using the Gauss mean-value theorem, assumed to be valid under these conditions.

EXERCISE 11 Prove the corresponding minimum principle.

EXERCISE 12 Using the minimum–maximum principle, prove the uniqueness of regular solutions of the Dirichlet problem for any Green region A in E^3, where the condition $u \in C^1(A \cup \delta A)$ is weakened to $u \in C(A \cup \delta A)$.

Of course, all the above can be accomplished with appropriate modification in E^n, $n \neq 3$. One of the modifications will be that the quantity 4π will be replaced by the surface area of a unit sphere in E^n, classically denoted by Ω_n. To derive the third identity from the second for $n = 2$ we use $v = \ln(r_{PQ})$; for $n \geq 3$ we use $v = (1/r_{PQ})^{n-1}$.

EXERCISE 13 Derive the third identity for $n = 2$.

EXERCISE 14 Derive the third identity for $n > 3$.

EXERCISE 15. Derive the Gauss mean-value theorem for $n = 2$; for $n \geq 3$.

EXERCISE 16 Prove the maximum principle for $n \geq 2$.

4 THE GREEN FUNCTION

The Green function is used to obtain integral representations for solutions to the Dirichlet problem for the Laplace and Poisson equations. It is thus akin to Riemann's function presented in Chapter 7. Both are known as "resolvent functions" in modern analysis. The Riemann function was undoubtedly constructed to perform the role of a Green function for hyperbolic type, but, unlike the Riemann function, the Green function forms a singular kernel.

Definition Let A be a region in E^n with boundary A, let $g: A \times A \to E^1$, and write $g = g(P, Q)$. Then g is said to be a Green function for the region A if

1. for $g(Q) = g(P_0, Q)$, then
    for every $\rho > 0$, $g \in C^2(A - S(\rho, P_0))$, $g \in C((A - S(\rho, P_0)) \cup \delta A)$;
    for $g(P) = g(P, Q_0)$, then
    for every $\rho > 0$, $g \in C^2(A - S(\rho, Q_0))$, $g \in C((A - S(\rho, Q_0)) \cup \delta A)$;
2. for $n = 2$,

$$g(P, Q) = h(P, Q) + \ln r_{PQ},$$

and for $n \geq 3$

$$g(P, Q) = h(P, Q) + (1/r_{PQ})^{n-2}$$

where

$$\text{for} \quad h(P_0, Q) = h(Q), \qquad \Delta h(Q) = 0 \quad \text{in} \quad A,$$

$$\text{for} \quad h(P, Q_0) = h(P), \qquad \Delta h(P) = 0 \quad \text{in} \quad A;$$

and

3. $P \in \delta A$ or $Q \in \delta A$ implies $g(P, Q) = 0$.

Briefly a Green function on A is the sum of a fundamental solution and a harmonic function, the sum vanishing on the boundary of A. Uniqueness is immediate because the difference of any two Green functions is harmonic and vanishes on the boundary, and we have already established uniqueness for the Dirichlet problem. The functions g themselves are not zero because, in fact, $g(P)$ is unbounded for P in any neighborhood of Q.

5 REPRESENTATION THEOREMS USING THE GREEN FUNCTION

We again utilize Fig. 47. Consider the Dirichlet problem for the Laplace equation

$$\Delta u = 0 \quad \text{in} \quad A \tag{10.5.1}$$

$$u = f \quad \text{in} \quad \delta A \tag{10.5.2}$$

and $u \in C^2(A)$, $u \in C^1(A \cup \delta A)$. Using the second Green identity much as we did to get the third, only this time with $v = g(P, Q)$ which is a function that is harmonic in $A - S$,

$$0 = \int_{A-S} (u \, \Delta g - g \, \Delta u) \, dA_Q = \int_{\delta A} (u \, \partial g/\partial n - g \, \partial u/\partial n) \, ds_Q$$

$$+ \int_{\delta S} (u \, \partial g/n - g \, \partial u/\partial n) \, ds_Q,$$

or, since $g = 0$ and $u = f$ in δA,

$$0 = \int_{\delta A} (f \, \partial g/\partial n) \, ds_Q + \int_{\delta S} (u \, \partial g/\partial n) \, ds_Q - \int_{\delta S} (g \, \partial u/\partial n) \, ds_Q. \tag{10.5.3}$$

The first integral will be left now as it is. In E^3 the second can be written for some $Q^* \in \delta S$, as

$$4\pi\rho^2 (\rho^{-2} + (\partial h(Q^*, P)/\partial n)) u(Q^*),$$

the limit of which as $\rho \to 0$ is

$$4\pi u(P), \tag{10.5.4}$$

and the third, by a similar argument vanishes with ρ. In total then, shrinking

$S(\rho, P)$ to a point, (10.5.3) becomes

$$u(P) = -(1/4\pi) \int_{\delta A} f \partial g / \partial n \, ds_Q, \tag{10.5.5}$$

the promised integral representation. It contributes almost nothing to the existence of a solution to (10.5.1), (10.5.2), the existence of a Green function and a solution to the Dirichlet problem being equivalent except for properties imposed on the boundary data f.

Stepping up from (10.5.1), now consider the problem

$$\Delta u = -4\pi\rho(P) \quad \text{in} \quad A \subset E^3, \tag{10.5.6}$$

$$u = f \quad \text{in} \quad \delta A, \tag{10.5.7}$$

and $u \in C^2(A)$, $u \in C^1(A \cup \delta A)$. For this purpose, let $u = u_1 + u_2$ where

$$\Delta u_1 = 0 \qquad \text{in} \quad A, \qquad u_1 = f \quad \text{in} \quad \delta A, \tag{10.5.8}$$

$$\Delta u_2 = -4\pi\rho(P) \quad \text{in} \quad A, \qquad u_2 = 0 \quad \text{in} \quad \delta A. \tag{10.5.9}$$

Of course, (10.5.5) gives the component u_1. Using the second identity in the same manner as before with $v = g(P, Q)$, but now with $u = u_2$, we have

$$4\pi \int_{A-S} (g\rho(P)) \, dA_Q = \int_S (u_2 \, \partial g / \partial n) \, ds_Q - \int_S (g \, \partial u_2 / \partial n) \, ds_Q,$$

the integrals on δA having vanished because both g and u_2 vanish there. By arguments now tendered several times, the first integral on the right goes to $4\pi u_2(P)$, and the second goes to zero, as the radius of S decreases. This gives

$$u_2(P) = \int_A (g\rho) \, dA_Q, \tag{10.5.10}$$

the total answer for (10.5.6), (10.5.7) being

$$u = u_1 + u_2 = \int_A (g\rho) \, dA_Q - \frac{1}{4\pi} \int_{\delta S} (f \, \partial g / \partial n) \, ds_Q. \tag{10.5.11}$$

The integral on the right of (10.5.10) [or even (10.5.5)] is appropriately referred to as the Green operator, since it maps a set of functions ρ defined on A (or f defined on δA) into the set of functions u on A. Since on a function space, an operator is just a function on the space into the space (see Part IV for this terminology), the integral in (10.5.10) can, albeit with some confusion, be referred to as a "Green function." The practice of so referring to it is confusing in our classical context, but this practice has become so popular among modern physicists that the original practice we follow here may seem confusing.

Really, in speaking of eigenvalues (as we did in Chapter 4) in this context one would do well to speak of "eigenvalues of the Green operator (10.5.10)" rather than of "eigenvalues of the Laplace equation with respect to the Dirichlet problem," but certainly one cannot simply speak of "eigenvalues of the Laplace equation." A differential equation alone does not have a resolvent function. The student may wish to consult Kellogg [24] to study Green functions (of the second kind), associated with the Neumann problem and integral representations of solutions. The eigenvalues of the attendant operators will be different than those of the Green operator (10.5.10) though they are all related to the same differential equation.

We will also leave it to the student to verify that all the above may be adapted to arguments in E^n, $n \geqslant 2$, wherever we have only proven our statements in E^3. The brief sketch in Chapter 3, Section 5, of numerical methods for the Dirichlet problem was regarded by some of the leaders of the last generation—we refer specifically to M. Picone and R. Courant—as a breaking away from the tight binding of the Green function. However, the Green function is not useless. We would rarely numerically solve a large system of linear algebraic equations by determinants or by inverting a matrix, but both theories are extremely useful as a basis for deductions; the latter is an excellent representation just as (10.5.5) and (10.5.10) are, but we would probably normally refuse to use any one of them to compute numerical answers.

6 VARIATIONAL METHODS

For many, perhaps even for most, applications of partial differential equations to physics, a detailed solution is not really required. Often what is of more practical interest is an integral over either some volume or some surface. This is evident in flow phenomena where, though a distribution of pressures and velocities over a flow field may be of some academic interest, it is often only certain integrals of pressure (giving drag or thrust) or of velocity [giving volume flow rate, see (4.7.9)] over the surface of a body (or chamber or control device) that determine the dynamical behavior of the system. When these entities can be expressed as positive-definite quadratic forms, it is often easier to obtain upper and lower bounds for them than it is to approximate the solutions from which they are fabricated. Here we give some "greatest lower bound–least upper bound" formulations for certain important quadratic forms that point the way toward approximation methods based on variational procedures. Applications to problems in elasticity, electrodynamics, and fluid mechanics, as well as a certain eigenvalue problem, will motivate our discussion.

7 DESCRIPTION OF TORSIONAL RIGIDITY

Consider an infinitely long cylinder of a homogeneous, isotropic medium with simply connected cross section R, this cylinder being subject to a pure torsion. The Saint-Venant theory says that the stresses inside the cylinder can be evaluated in terms of the stress function φ, which is a solution of the Dirichlet problem,

$$\Delta\varphi = -2 \quad \text{in} \quad R, \qquad \varphi = 0 \quad \text{in} \quad \delta R, \qquad (10.7.1)$$

where δR is the boundary of R. For many applications, it is sufficient to obtain the torque per unit angle of twist, the *torsional rigidity*, which is proportional to the Dirichlet integral,

$$D(\varphi) = \iint_R (\varphi_x{}^2 + \varphi_y{}^2)\, dR. \qquad (10.7.2)$$

Since the differential equation and boundary function (10.7.1) are known, and they determine the function φ, the number $D(\varphi)$ may be thought of as depending only on the domain R. Such a number is often referred to as a domain functional. In general an operator giving a single number from a function is called a functional, or, in modern context, we think of the functions as members of a linear vector space V over the real numbers R^1, and then a functional is a function on the elements of the space into its field of scalars R^1.[†]

A linear functional is such a function L with the property

$$L(\alpha x + \beta y) = \alpha L(x) + \beta L(y),$$

where α and β are real numbers while x and y are functions (elements of the vector space). A bilinear form is a function B on $V \times V$ into R^1, which we may regard as a real-valued function of two function variables, $B = B(x, y)$, and is such that it is linear in one variable when the other is "held fixed." That is, we have

$$B(\alpha x_1 + \beta x_2, y) = \alpha B(x_1, y) + \beta B(x_2, y) \qquad (10.7.3)$$

and

$$B(x, \alpha y_1 + \beta y_2) = \alpha B(x, y_1) + \beta B(x, y_2). \qquad (10.7.4)$$

The functional $Q: V \to R^1$ is said to be a quadratic functional if there is a bilinear form B such that

$$Q(x) = B(x, x). \qquad (10.7.5)$$

---

† See Part IV for details of this terminology.

It is said to be positive definite if $x \neq 0$ implies $Q(x) > 0$; positive semidefinite if $x \neq 0$ implies $Q(x) \geq 0$. It should be noted that by this definition $Q(0) = 0$.

It will be found that the Dirichlet integral (10.7.2) is a quadratic functional. Other quantities of physical interest, similar to that defined in (10.7.1), (10.7.2) in that they are quadratic functionals of solutions to the Laplace or Poisson equation with given boundary conditions, will now be given.

8 DESCRIPTION OF ELECTROSTATIC CAPACITANCE, POLARIZATION, AND VIRTUAL MASS

Where R is a region in E^3 (or possibly E^2) exterior to its boundary δR, the *capacity* of the boundary surface δR is given by

$$C = (1/4\pi)D(\varphi) \qquad (10.8.1)$$

where

$$\Delta\varphi = 0 \quad \text{in} \quad R, \qquad \varphi = 1 \quad \text{in} \quad \delta R, \qquad \varphi = 0 \quad \text{at} \quad \infty, \qquad (10.8.2)$$

and $D(\varphi)$ is the Dirichlet integral (10.7.2), now taken over an infinite region R.

The *polarization in the x direction* of an infinite electric field in a region R exterior to its boundary δR and due to the presence of the conducting surface δR is again given by $D(\varphi)$ where

$$\Delta\varphi = 0 \quad \text{in} \quad R, \qquad \varphi = x + \text{constant} \quad \text{in} \quad \delta R,$$

$$\varphi = O(1/r^2) \quad \text{at} \quad \infty. \qquad (10.8.3)$$

In hydrodynamics the *virtual mass* is regarded as the effective mass added to account for kinetic energy. The *virtual mass* of a body inside a surface δR with exterior R due to an incompressible flow in the x direction is proportional to $D(\psi)$ with

$$\Delta\psi = 0 \quad \text{in} \quad R, \qquad \partial\psi/\partial n = \partial x/\partial n \quad \text{in} \quad \delta R,$$

$$\psi = O(1/r^2) \quad \text{at} \quad \infty. \qquad (10.8.4)$$

Further treatments of electrostatic capacitance, polarization, and virtual mass can be found in [36].

For capacitance, polarization, and virtual mass the boundary-value problems are associated with exterior regions and not with interior regions, as was the case for torsional rigidity.

9 THE DIRICHLET INTEGRAL AS A QUADRATIC FUNCTIONAL

The Dirichlet integral

$$D(f) = \int_R |\operatorname{grad} f|^2 \, dR \qquad (10.9.1)$$

is a positive-semidefinite quadratic functional (see Section 7). The bilinear form B associated with any quadratic functional Q is given by

$$B(f,g) = \tfrac{1}{4}[Q(f+g) - Q(f-g)]. \qquad (10.9.2)$$

If $Q(f) \geqslant 0$, then for all real numbers α,

$$Q(f+\alpha g) = Q(f) + 2\alpha B(f,g) + \alpha^2 Q(g) \geqslant 0$$

or, for $\alpha = -1$,

$$2B(f,g) \leqslant Q(f) + Q(g)$$

or

$$B^2(f,g) \leqslant \tfrac{1}{2}Q^2(f) + Q(f)Q(g) + \tfrac{1}{2}Q^2(g)$$

or

$$B^2(f,g) \leqslant Q(f)Q(g). \qquad (10.9.3)$$

The latter will be recognized as the famous inequality variously attributed to Cauchy, Schwarz, or Bunyokovski, depending on at which level of generality one stops considering a contribution original and which nationality he feels like favoring. The modern tendency is to call (10.9.3) the CBS inequality and thus avoid commitment. It is, in this general form, the principle tool in many variational methods, and many of the classical maximum and minimum principles follow from it [7].

Let us confine ourselves for the moment to a simply connected region R interior to a boundary δR and let

$$\Delta\varphi = 0 \quad \text{in} \quad R, \qquad \varphi = f \quad \text{in} \quad \delta R. \qquad (10.9.4)$$

If v is *any* function for which the Dirichlet integral exists, then by the CBS inequality,

$$D(\varphi, v)^2 \leqslant D(\varphi) D(v), \qquad (10.9.5)$$

where $D(\varphi, v)$ is the bilinear form associated with the Dirichlet integral $D(\varphi)$ or $D(v)$. Further, if $v \in C^1(R)$ and $v = f$ on δR, then by Green's first identity,

$$D(\varphi, v) = \int_R (\operatorname{grad} \varphi, \operatorname{grad} v) \, dR = -\int_R v \, \Delta\varphi \, dR + \int_{\delta R} v \, \partial\varphi/\partial n \, ds \qquad (10.9.6)$$

or

$$D(\varphi, v) = \int_{\delta R} \varphi \, \partial\varphi/\partial n \, ds = D(\varphi).$$

Thus, for such functions φ, by (10.9.5) we have $D(\varphi) \leqslant D(v)$, and this is the Dirichlet principle,

$$D(\varphi) = \min_{v = f \text{ in } \delta R} D(v). \tag{10.9.7}$$

Next, let v satisfy $v \in C^2(R)$ and let $\Delta v = 0$ in R (instead of $v = f$ in δR). Then by Green's theorem

$$D(\varphi, v) = \int_R (\text{grad } v, \text{grad } \varphi) \, dR = -\int_R \varphi \, \Delta v \, dR + \int_{\delta R} \varphi \, \partial v/\partial n \, ds$$

$$= \int_{\delta R} f \, \partial v/\partial n \, ds, \tag{10.9.8}$$

or, from (10.9.5),

$$D(\varphi) \geqslant \frac{(\int f \, \partial v/\partial n \, ds)^2}{D(v)},$$

or, since equality is attained for $v = \varphi$,

$$D(\varphi) = \max_{\Delta v = 0} \frac{(\int f \, \partial v/\partial n \, ds)^2}{D(v)}. \tag{10.9.9}$$

This is Sir William Thompson's (Lord Kelvin's) principle.

The Dirichlet and Thompson principles are both still valid for exterior regions R, providing φ and v are such that the boundary terms vanish at infinity. That is, for the interior region between a sphere S_r of radius r and the surface δR, the boundary terms of (10.9.6) and (10.9.8) become

$$-\int_{S_r} (v \, \partial\varphi/\partial n) \, ds + \int_{\delta R} (v \, \partial\varphi/\partial n) \, ds$$

and

$$-\int_{S_r} (\varphi \, \partial v/\partial n) \, ds + \int_{\delta R} (\varphi \, \partial v/\partial n) \, ds,$$

respectively. Then (10.9.7) and (10.9.9) apply unchanged to the exterior region of δR provided that

$$\lim_{r \to \infty} \int_{S_r} v \, \partial\varphi/\partial n \, ds = \lim_{r \to \infty} \int_{S_r} \varphi \, \partial v/\partial n \, ds = 0,$$

and this will be the case if both φ and v are $O(1/r)$.

10 DIRICHLET AND THOMPSON PRINCIPLES FOR SOME PHYSICAL ENTITIES

The fact that f is determined only to within a constant in the polarization problem of Section 8 is unimportant since from Green's theorem, as above using a Neumann condition,

$$0 = - \int_{S_r} \partial v / \partial n \, ds + \int_{\delta R} \partial v / \partial n \, ds,$$

and, in the limit,

$$0 = \int_{\delta R} \partial v / \partial n \, ds. \tag{10.10.1}$$

That is, if we require $v = O(1/r^2)$, (10.9.9) is unaffected by the arbitrariness involved in f. We thus have both a maximum and minimum principle for capacity and polarization.

The same is true for torsional rigidity and virtual mass. Torsional rigidity, of course, is an interior problem. That it can be reduced to the problem (10.9.4) for the homogeneous differential equation was noted by Diaz and Weinstein [8] as follows:

$$\Delta \varphi = -2 \Rightarrow \Delta(\varphi + \tfrac{1}{2}r^2) = 0; \tag{10.10.2}$$

then expanding $D(\varphi + \tfrac{1}{2}r^2)$, applying the first Green identity and using the fact that φ vanishes on the boundary, one has

$$D(\varphi + \tfrac{1}{2}r^2) = D(\varphi) + D(\varphi, r^2) + \tfrac{1}{4}D(r^2)$$

$$= D(\varphi) - 4 \int_R \varphi \, dR + \int_R r^2 \, dR$$

$$= D(\varphi) - 2D(\varphi) + \int_R r^2 \, dR$$

or

$$D(\varphi) = \int_R r^2 \, dR - D(\varphi + \tfrac{1}{2}r^2). \tag{10.10.3}$$

Noting that $\int_R r^2 \, dR$ is just the polar moment of inertia of R, one sees that it is sufficient to approximate $D(\varphi + \tfrac{1}{2}r^2)$ in order to find $D(\varphi)$.

For virtual mass, consider first the general interior Neumann problem,

$$\Delta \psi = 0 \quad \text{in} \quad R, \qquad \partial \psi / \partial n = g \quad \text{in} \quad \delta R, \tag{10.10.4}$$

Here

$$D(v, \psi) = \int_{\delta R} (v \, \partial\psi/\partial n) \, ds = \int_{\delta R} (vg) \, ds,$$

and, from the CBS inequality,

$$D(\psi) \geqslant \frac{(\int_{\delta R} vg \, ds)^2}{D(v)}, \qquad (10.10.5)$$

where v need only be continuous. On the other hand, let

$$\Delta v = 0 \quad \text{in} \quad R, \qquad \partial v/\partial n = g \quad \text{in} \quad \delta R. \qquad (10.10.6)$$

Then

$$D(v, \psi) = \int_{\delta R} \psi \, \partial v/\partial n \, ds = \int_{\delta R} \psi \, \partial\psi/\partial n \, ds = D(\psi)$$

so that

$$D(\psi) \leqslant D(v),$$

and, as before, for the exterior problem, it is only required that v, as well as ψ, be $O(1/r^2)$ at infinity.

11 EIGENVALUES AS QUADRATIC FUNCTIONALS

A significantly different type of domain functional associated with quadratic functionals are eigenvalues. Consider a problem of the form

$$\Delta u + \lambda u = 0 \quad \text{in} \quad R, \qquad u = 0 \quad \text{in} \quad \delta R, \qquad (10.11.1)$$

and consider only regular solutions. This boundary value problem has nonzero solutions only for a discrete set of values of the constant λ. These values are denoted by $\lambda_1, \lambda_2, \ldots$. The situation is much like the problem of finding eigenvalues for a matrix A, where one is asked to find nonzero solutions of the equation $Ax + \lambda x = 0$, only for that case the set of eigenvalues is a finite set while here we obtain an infinite sequence of eigenvalues.

Multiplying (10.11.1) by u, we have

$$-\sum_{i=1}^{n} u_{x_i}^2 + \sum_{i=1}^{n} (uu_{x_i})_{x_i} + \lambda u^2 = \sum_{i=1}^{n} uu_{x_i x_i} + \lambda u^2 = 0,$$

or, from the homogeneous boundary conditions and the fact that $u \in C^1(R \cup \delta R)$,

$$-D(u) + \lambda \int_R u^2 \, dR = 0,$$

175

so that

$$\lambda = \frac{D(u)}{\int_R u^2 \, dR}.$$ (10.11.2)

Thus we see, at least, that $\lambda_i > 0$ for every $i = 1, 2, \dots$.

It can be shown that $\lambda_i \to \infty$, and the eigenvalues can be ordered,

$$\lambda_1 \leqslant \lambda_2 \leqslant \cdots .$$

From (10.11.2), then,

$$\lambda_1 = \min_{u = 0 \text{ in } \delta R} \frac{D(u)}{\int_R u^2 \, dR},$$

and, if u_i is the eigenfunction (solution) corresponding to λ_i,

$$\lambda_{i+1} = \min_{u = 0 \text{ in } \delta R} \left(\frac{D(u)}{\int_R u^2 \, dR} \right),$$ (10.11.3)

where

$$\int_R (u u_i) \, dR = 0 \qquad \text{for} \quad i = 1, \dots, n.$$

Equation (10.11.3) is known as the Rayleigh–Ritz principle. It and other related considerations which allow for computations of upper bounds of eigenvalues were discovered first by Lord Rayleigh and independently, 2 years later, by W. Ritz. The principle (10.11.3) was used earlier in Chapter 4 to prove mean-square asymptotic uniqueness of viscous flow in a rigid pipe. Work still continues toward finding good lower bounds. We recommend works of Fichera [11, 12], Weinberger [43], and Weinstein [44, 45]. The methods of Weinstein for finding lower bounds motivated the works of a generation of mathematicians and engineers [18] in this area, helped to establish the usefulness of Hilbert space theory (and, therefore, indirectly of functional analysis in general) when it was in its infancy, and still finds a usefulness in modern applications [2]. The works of the authors quoted can be said to represent different "schools," and each will find enthusiastic proponents in any group of mathematical scientists.

11

Solution of the Cauchy Problem for the Wave Equation in Terms of Retarded Potentials

1 INTRODUCTION

We are interested in the Cauchy problem

$$u_{xx} + u_{yy} + u_{zz} - (1/a^2)u_{tt} = F(x, y, z, t) \qquad (11.1.1)$$

$$u(x, y, z, 0) = f(x, y, z), \qquad u_t(x, y, z, 0) = g(x, y, z) \qquad (11.1.2)$$

where $(x, y, z) \in E^3$, $t \geqslant 0$, and $a > 0$. We reserve the right to think of f and g defined only on some set A in E^3, perhaps bounded, but this will play no role for the present. Physically, one may think of a vibrating elastic solid or of Maxwell's field equations where u is the scalar potential and $F = -\rho/\varepsilon$. Here ρ is charge density and ε is permittivity [32]. To obtain an a priori representation theorem we develop a generalization of the third Green identity called the Kirchhoff formula in which the E^4 wave operator, rather than the Laplace operator, appears in the volume integral. In the former case, we obtained a solution of the Laplace equation in terms of the values of the function on the boundary of a region *and* its normal derivative there, but we showed by way of a uniqueness proof and a Green function argument, that in order for the solution to be valid in the entire region, the two functions on the boundary could not be independently prescribed. One was, in fact, completely determined by the other (or the other and a value at one point). Thus the a priori formula did not, in fact, describe a solution in any general case. Here, on the other hand, it can be shown, simply by differentiating the expression obtained, that the a priori formula, properly reworked, does indeed provide the solution (if $f \in C^3$, $g \in C^2$, and $F \in C^2$), and, moreover, one notes that it supplies a uniqueness proof. The solution so obtained is said to be in terms of retarded potentials for reasons that will appear from its form. The volume term is known as the Poisson term, and the solution is sometimes referred to as the Poisson solution of the wave equation in terms of retarded potentials. The concept of retarded potentials can be made to play a basic role in electrodynamics, as for example, in [32]. The student who at this point is looking for ancillary reading will find this text rewarding. The entire setting here, especially the work of Kirchhoff, was originally quite thoroughly embedded in electrodynamics.

2 KIRCHHOFF'S FORMULA

Let

$$t^* = t - r/a \tag{11.2.1}$$

where $a \in R^1$ and $a > 0$, and let $u(x, y, z, t^*)$ be defined for $(x, y, z) \in E^3$ and for $t^* \geqslant 0$ (i.e., $t \geqslant r/a$). Let φ be the function at a time r/a in the past; i.e., let

$$\varphi(x, y, z, t) = u(x, y, z, t^*) \in C^2 \tag{11.2.2}$$

for all $(x, y, z) \in E^3$ and for $t \geqslant r/a$ (i.e., in the forward conoid, see Fig. 49). We have

$$t_x^* = -(x - x_0)/ar, \qquad t_y^* = -(y - y_0)/ar, \qquad t_z^* = -(z - z_0)/ar$$

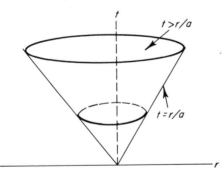

FIG. 49 The forward conoid. The inside is given by $t > r/a$.

178

and

$$\varphi_x = u_x + u_{t*}\, t_x^{\;*} = u_x - \frac{(x-x_0)}{ar}\, u_{t*},$$

$$\varphi_y = u_y + u_{t*}\, t_y^{\;*} = u_y - \frac{(y-y_0)}{ar}\, u_{t*},$$

$$\varphi_z = u_z + u_{t*}\, t_z^{\;*} = u_z - \frac{(z-z_0)}{ar}\, u_{t*},$$

and

$$\varphi_{xx} = \left[u_{xx} + u_{t*t*}\left(-\frac{(x-x_0)}{ar} \right) \right] - \frac{(x-x_0)}{ar}\left[u_{t*x} + u_{t*t*}\left(-\frac{(x-x_0)}{ar} \right) \right]$$

$$+ \left[-\frac{1}{ar} + \frac{(x-x_0)^2}{ar^3} \right] u_{t*}$$

$$= u_{xx} - 2\frac{(x-x_0)}{ar}\, u_{t*x} + \frac{(x-x_0)^2}{a^2 r^2}\, u_{t*t*} + \left[-\frac{1}{ar} + \frac{(x-x_0)^2}{ar^3} \right] u_{t*},$$

$$\varphi_{yy} = u_{yy} - 2\frac{(y-y_0)}{ar}\, u_{t*y} + \frac{(y-y_0)^2}{a^2 r^2}\, u_{t*t*} + \left[-\frac{1}{ar} + \frac{(y-y_0)^2}{ar^3} \right] u_{t*},$$

$$\varphi_{zz} = u_{zz} - 2\frac{(z-z_0)}{ar}\, u_{t*y} + \frac{(z-z_0)^2}{a^2 r^2}\, u_{t*t*} + \left[-\frac{1}{ar} + \frac{(z-z_0)^2}{ar^3} \right] u_{t*},$$

so that

$$\Delta\varphi = \Delta u + \frac{1}{a^2}\, u_{t*t*} - \frac{2}{a}\left\{ \frac{(x-x_0)}{r}\, u_{xt*} + \frac{(y-y_0)}{r}\, u_{yt*} + \frac{(z-z_0)}{r}\, u_{zt*} \right\} - \frac{2}{ar}\, u_{t*}.$$

$$(11.2.3)$$

We now replace the braces in (11.2.3): Because $(r/a)_t = 0$,

$$\varphi_t(x,y,z,t) = u_t(x,y,z,t-r/a) = u_{t*}(x,y,z,t-r/a),$$

and

$$\varphi_{tx} = u_{t*x} + u_{t*t*}\, t_x^{\;*} = u_{t*x} - \frac{(x-x_0)}{ar}\, u_{t*t*}$$

$$\varphi_{ty} = u_{t*y} + u_{t*t*}\, t_y^{\;*} = u_{t*y} - \frac{(y-y_0)}{ar}\, u_{t*t*}$$

$$\varphi_{tz} = u_{t*z} + u_{t*t*}\, t_z^{\;*} = u_{t*z} - \frac{(z-z_0)}{ar}\, u_{t*t*},$$

and

$$-\frac{2}{a}\left\{\frac{(x-x_0)}{r}\varphi_{tx} + \frac{(y-y_0)}{r}\varphi_{ty} + \frac{(z-z_0)}{r}\varphi_{tz}\right\}$$

$$= -\frac{2}{a}\left\{\frac{(x-x_0)}{r}u_{t*x} + \frac{(y-y_0)}{r}u_{t*y} + \frac{(z-z_0)}{r}u_{t*z}\right\} + \frac{2}{a^2}u_{t*t*}$$

or

$$-\frac{2}{a}\left\{\frac{(x-x_0)}{r}u_{t*x} + \frac{(y-y_0)}{r}u_{t*y} + \frac{(z-z_0)}{r}u_{t*z}\right\}$$

$$= -\frac{2}{a}\left\{\frac{(x-x_0)}{r}\varphi_{xx} + \frac{(y-y_0)}{r}\varphi_{xy} + \frac{(z-z_0)}{r}\varphi_{tz}\right\} - \frac{2}{a^2}u_{t*t*}.$$

Therefore, from (11.2.3), we have

$$\frac{1}{r}\Delta\varphi = \frac{1}{r}\left(\Delta u - \frac{1}{a^2}u_{t*t*}\right)$$

$$-\frac{2}{a}\left\{\frac{(x-x_0)}{r^2}\varphi_{tx} + \frac{(y-y_0)}{r^2}\varphi_{ty} + \frac{(z-z_0)}{r^2}\varphi_{tz}\right\} - \frac{2}{ar^2}\varphi_t.$$

$$\tag{11.2.4}$$

But

$$\left(\frac{x-x_0}{r}\right)_x + \left(\frac{y-y_0}{r}\right)_y + \left(\frac{z-z_0}{r}\right)_z = \frac{1}{r^2},$$

so we have

$$\frac{1}{r}\Delta\varphi = \frac{1}{r}\left(\Delta u - \frac{1}{a^2}u_{t*t*}\right) - \frac{2}{a}\left\{\left(\frac{x-x_0}{r^2}\varphi_t\right)_x + \left(\frac{y-y_0}{r^2}\varphi_t\right)_y + \left(\frac{z-z_0}{r^2}\varphi_t\right)_z\right\}$$

$$= \frac{1}{r}\left(\Delta u - \frac{1}{a^2}u_{t*t*}\right) = -\frac{2}{a}\operatorname{div}(\varphi_t \cdot \operatorname{grad}\ln r). \tag{11.2.5}$$

We are now prepared to use the second Green identity (5.6.7) with $u = 1/r_{PQ}$ and $v = \varphi$ [the retarded function u as defined by (11.2.2)] on a region $A - (s(\rho, P) \cup \delta S(\rho, P))$ for some region $A \subset E^3$ with boundary δA (see Fig. 47). Of course, we assume that $P, Q \in A$ and we assume that $\varphi: A \cup \delta A \to R^1$, $\varphi \in C^2(A)$, $C^1(A \cup \delta A)$. For short we write $A - S$ for $A - (S(\rho, P) \cup$

$\delta S(\rho, P))$ and $1/r$ for $1/r_{PQ}$; then

$$\int_{A-S} ((1/r)\,\Delta\varphi)\,dA_Q$$

$$= \int_{\delta S} ((1/r)\,\varphi_n - \varphi(1/r)_n)\,ds_Q + \int_{\delta A} ((1/r)\,\varphi_n - \varphi(1/r)_n)\,ds_Q. \qquad (11.2.6)$$

We consider the limit of each of the integrals in (11.2.6) in turn:

(1) From (11.2.5) the first integral may be written

$$\int_{A-S} \left(\Delta u - \frac{1}{a^2}\,u_{t^*t^*}\right)\frac{1}{r}\,dA_Q - \frac{2}{a}\int_{A-S} \operatorname{div}\left(\varphi_t\,\text{grad}\,\ln\frac{1}{r}\right)dA_Q,$$

the latter term of which can be written as

$$-\frac{2}{a}\int_{\delta A + \delta S} \left(\bar{n}, \varphi_t\,\text{grad}\,\ln\frac{1}{r}\right)ds_Q$$

$$= -\frac{2}{a}\int_{\delta A} \frac{1}{r}\frac{\partial r}{\partial n}\,\varphi_t\,ds_Q$$

$$+ \int_0^{2\pi} d\Phi \int_0^{\pi} \left\{\frac{x-x_0}{\rho_2}\,\varphi_t\left(-\frac{(x-x_0)}{\rho}\right) + \frac{y-y_0}{\rho^2}\,\varphi_t\left(-\frac{(y-y_0)}{\rho}\right)\right.$$

$$\left. + \frac{z-z_0}{\rho^2}\,\varphi_t\left(-\frac{(z-z_0)}{\rho}\right)\right\}\rho^2 \sin\theta\,d\theta,$$

where we have introduced coordinates (r, θ, Φ), $0 \leqslant \Phi \leqslant 2\pi$, $0 \leqslant \theta \leqslant \pi$, in the integral over δS and noted that there, $\partial/\partial n = -\partial/\partial r$. The last integral above can be written

$$\int_0^{2\pi} d\Phi \int_0^{\pi} -\left\{\frac{(x-x_0)^2 + (y-y_0)^2 + (y-y_0)^2 + (z-z_0)^2}{\rho^3}\right\}\varphi_t \rho^2 \sin\theta\,d\theta$$

or

$$\int_0^{2\pi} d\Phi \int_0^{\pi} (-1/\rho)\,\varphi_t \rho^2 \sin\theta\,d\theta$$

so its limit as $\rho \to 0$ is zero.

(2) The integral in (11.2.6) over δS can be written as

$$\int_0^{2\pi} d\Phi \int_0^{\pi} ((-1/\rho)\,\varphi_r - \varphi(1/\rho^2))\rho^2 \sin\theta\,d\theta$$

181

or

$$\rho \int_0^{2\pi} d\Phi \int_0^{\pi} (-\varphi_r) \sin \theta \, d\theta + \int_0^{2\pi} d\Phi \int_0^{\pi} (-\varphi) \sin \theta \, d\theta.$$

Since the first term goes to zero with ρ, we have that the limit as $\rho \to 0$ is

$$-\lim_{\rho \to 0} \varphi(Q^*) \int_0^{2\pi} d\Phi \int_0^{\pi} \sin \theta \, d\theta,$$

where Q^* is any point on $\delta S(\rho, P)$, and this limit is $-4\pi\varphi(x_0, y_0, z_0, t)$.

(3) The third integral in (11.2.6) is left as it is written.

In total, we now have

$$\int_A \frac{1}{r} \left(\Delta u - \frac{1}{a^2} u_{t^*t^*} \right) dA_Q - \frac{2}{a} \int_{\delta A} \left(\frac{1}{r} \frac{\partial r}{\partial n} \varphi_t \right) ds_Q$$

$$= \int_{\delta A} \left[\frac{1}{r} \frac{\partial \varphi}{\partial n} - \varphi \frac{\partial}{\partial n} \left(\frac{1}{r} \right) \right] ds_Q - 4\pi\varphi(P, t). \qquad (11.2.7)$$

If $P \subset \mathscr{C}(A \cup \delta A)$, the integral in (11.2.6) over δS does not occur, and thus the last term in (11.2.7) vanishes. Noting that $r = r_{PQ} = 0$ at $P = Q$, so that $\varphi(P, t) = \varphi(x_0, y_0, z_0, t) = u(x_0, y_0, z_0, t) = u(P, t)$, we are ready to summarize.

Theorem Where φ is the retarded value (11.2.2) of u and where A is a region in E^3, for every $u \in C^2(A)$, $C^1(A \cup \delta A)$, $P \in A$ implies that u equals the following while $P \in \mathscr{C}A$ implies that 0 equals the following:

$$\frac{1}{4\pi} \left\{ \int_{\delta A} \frac{1}{r} \varphi_n \, ds_Q - \int_{\delta A} \varphi \left(\frac{1}{r} \right)_n ds_Q + \frac{2}{a} \int_{\delta A} \frac{1}{r} r_n \varphi_t \, ds_Q \right.$$

$$\left. - \int_A \frac{1}{r} \left(\Delta u - \frac{1}{a^2} u_{t^*t^*} \right) dA_Q \right\} \qquad (11.2.8)$$

$$= \frac{1}{4\pi} \left\{ \int_{\delta A} \frac{1}{r} v_n \, ds_Q - \int_{\delta A} \varphi \left(\frac{1}{r} \right)_n ds_Q + \frac{1}{a} \int_{\delta A} \frac{1}{r} r_n \varphi_t \, ds_Q \right.$$

$$\left. - \int_A \frac{1}{r} \left(\Delta u - \frac{1}{a^2} u_{t^*t^*} \right) dA_Q \right\}. \qquad (11.2.9)$$

The latter equality comes from

$$\varphi_x = u_x - \frac{x - x_0}{ar} u_{t^*}, \qquad u_n - \frac{1}{a} \frac{\partial r}{\partial n} u_{t^*} = \varphi_n$$

and

$$\frac{1}{r}\varphi_n = \frac{1}{r}u_n - \frac{1}{ar}\frac{\partial r}{\partial n}u_{t*}.$$

Equation (11.2.8) is the Kirchhoff identity. All terms except the time derivative term can be regarded as potentials. Since φ is said to be a retarded value of u, the first two integrals on the right are called retarded potentials.

3 SOLUTION OF THE CAUCHY PROBLEM

It remains to use the Kirchhoff identity (11.2.8), (11.2.9) for the purpose of solving the Cauchy problem (11.1.1), (11.1.2). In (11.2.8) we let

$$A = S(r, P), \qquad (11.3.1)$$

and, noting that now[†] $\partial/\partial n = \partial/\partial r$, we have

$$u(P, t) = \frac{1}{4\pi}\left\{\int_{\delta S}\frac{1}{r}u_n\,ds + \int_{\delta S}\varphi\frac{1}{r^2}\,ds + \frac{1}{a}\int_{\delta S}\frac{1}{r}\varphi_t\,ds\right.$$
$$\left. - \int_S\frac{1}{r}\left(\Delta u - \frac{1}{a^2}u_{t*t*}\right)ds\right\}. \qquad (11.3.2)$$

We have now dropped the Q subscript because it is clear that P is at the center of the sphere S and the integration variable is in S or δS.

If now we let (see Fig. 50)

$$r = at \qquad \text{so that} \qquad A = S(at, P) \qquad (11.3.3)$$

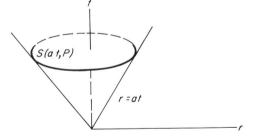

FIG. 50 The sphere (in E^3) with center P and radius at is pictured in a time-versus-radius plot.

[†] Earlier where the region was $A - S$, the outer normal on the δS portion of the boundary $\delta A \cup \delta S$ was $-r$, now when the region is S the outer normal is r.

and use (11.1.1), (11.1.2) so that

$$\varphi(x, y, z, r/a) = u(x, y, z, 0) = f(x, y, z) \tag{11.3.4}$$

and

$$\varphi_t(x, y, z, r/a) = u_t(x, y, z, 0) = g(x, y, z), \tag{11.3.5}$$

then (11.3.2) becomes

$$u(P, t) = \frac{1}{4\pi} \left\{ \frac{1}{at} \int_{\delta S} f_n \, ds + \frac{1}{a^2 t^2} \int_{\delta S} f \, ds + \frac{1}{a^2 t} \int_{\delta S} g \, ds \right.$$

$$\left. - \int_S \frac{1}{r} F\left(x, y, z, t - \frac{r}{a} \right) dA \right\} \tag{11.3.6}$$

provided f, g, and F are sufficiently smooth that all integrals listed exist.

This is a representation of our solution, an obvious fulfillment of the role intended by Kirchhoff for his form of the Green identity. We now proceed to write this representation in a more convenient form. For this purpose, we note that

$$\frac{\partial}{\partial t} \int_{\delta S} f \, ds = \frac{\partial}{\partial t} \int_0^{2\pi} d\Phi \int_0^\pi f(x_0 + at \sin \theta \cos \Phi,$$

$$y_0 + at \sin \theta \cos \Phi, z_0 + at \cos \theta)(a^2 t^2 \sin \theta) \, d\theta$$

$$= \int_2^{2\pi} d\Phi \int_0^\pi [f_x(a \sin \theta \cos \Phi) + f_y(a \sin \theta \sin \Phi) + f_z(a \cos \theta)]$$

$$\times (a^2 t^2 \sin \theta) \, d\theta + 2 \int_0^{2\pi} d\Phi \int_0^\pi f(a^2 t \sin \theta) \, d\theta$$

$$= a \int_0^{2\pi} d\Phi \int_0^\pi \left[f_x \frac{(x - x_0)}{at} + f_y \frac{(y - y_0)}{at} + f_z \frac{(z - z_0)}{at} \right] (a^2 t^2 \sin \theta) \, d\theta$$

$$+ \frac{2}{t} \int_0^{2\pi} d\Phi \int_0^\pi f(a^2 t^2 \sin \theta) \, d\theta$$

$$= a \int_{\delta S} f_n \, ds + \frac{2}{t} \int_{\delta S} f \, ds,$$

or we may write

$$\frac{1}{at} \int_{\delta S} f_n \, ds = \frac{1}{a^2 t} \frac{\partial}{\partial t} \int_{\delta S} f \, ds - \frac{2}{a^2 t} \int_{\delta S} f \, ds. \tag{11.3.7}$$

This then allows our a priori representation (11.3.6) to be written in the form

$$u(P, t) = \frac{1}{4\pi} \left\{ \frac{\partial}{\partial t} \left(\frac{1}{a^2 t} \int_{\delta S} f \, ds \right) + \frac{1}{a^2 t} \int_{\delta S} g \, ds - \int_{S} \frac{1}{r} F\left(x, y, z, t - \frac{r}{a} \right) dA \right\},$$

(11.3.8)

where, of course, it is to be remembered that $S = S(at, P)$.

For all the terms in our representation (11.3.8) to exist we need only require that $f \in C^1$, $g \in C$, $F \in C$, or less, but for this representation to satisfy the wave equation we require that

$$f \in C^3, \qquad g \in C^2, \qquad F \in C^2.$$

(11.3.9)

That it does then satisfy the differential equation (11.1.1) and the boundary conditions (11.1.2) will be demonstrated in the following sections, so the a priori representation is then found to be a solution. *It will be particularly in evidence in* Part IV *that the role of a priori inequalities in modern analysis is to establish existence, but rarely can they be used so directly for this purpose.* One should also note that the a priori representation here gives uniqueness immediately, and this is usually so.

We note once again, as we did at the beginning of this chapter, that the third Green identity (without the Kirchhoff modification) gives an a priori representation of the Dirichlet problem for the Poisson equation which is entirely analogous to (11.3.8), but that this representation proves not to be a solution unless u and u_n are very specially related on the boundary. The very specific nature of the relationship is revealed by solving the Dirichlet problem using a Green function or (since the Green function solution represents an a posteriori argument) by looking at the solution of special problems such as those we have selected for presentation in Chapter 3. Surely there is a lesson in this, relative to the question of static versus dynamic models raised in Chapter 4, if we could only learn to read it.

4 THE SOLUTION IN MEAN-VALUE FORM

The mean value of f at $P = (x_0, y_0, z_0)$ over the surface of a sphere of radius ρ is denoted by

$$\bar{f}(x_0, y_0, z_0, \rho) = (1/4\pi\rho^2) \int_{\delta S} f \, ds.$$

(11.4.1)

In spherical coordinates,

$$\bar{f}(P, \rho) = (1/4\pi\rho^2) \int_0^{2\pi} d\Phi \int_0^{\pi} f(x_0 + \rho \sin\theta \cos\varphi, \, y_0 + \rho \sin\theta \sin\varphi,$$

$$z_0 + \rho \cos\theta)(\rho^2 \sin\theta) \, d\theta$$

$$= (1/4\pi) \int_\omega f(x_0 + \rho x_1, \, y_0 + \rho\alpha_2, \, z_0 + \rho\alpha_3) \, d\omega, \tag{11.4.2}$$

where ω is the surface of the unit sphere $S(1, P)$ and $\alpha_1, \alpha_2, \alpha_3$ are variables of integration on that sphere expressed in rectangular coordinates. We have the occasion in what follows to use the notation (11.4.2) frequently. As a shorthand notation for (11.4.1) or (11.4.2) we write for $S = S(at, P)$,

$$\bar{f}(at) = \bar{f}(P, at) = (1/4\pi a^2 t^2) \int_{\delta S} f(x, y, z) \, ds.$$

Then with $F = 0$ our a priori representation (11.3.8) may be written

$$u(P, t) = \frac{1}{4\pi} \left\{ \frac{\partial}{\partial t} \left(t\bar{f}(at) + t\bar{g}(at) \right) \right\}, \tag{11.4.3}$$

which can be regarded as a generalization of the D'Alembert solution (see Chapter 1) to E^3, although how it reduces to that simple form required in E^2 is perhaps not immediately evident (see Hadamard's method of descent, Section 8).

5 VERIFICATION OF THE SOLUTION TO THE HOMOGENEOUS WAVE EQUATION

We now verify that (11.3.8) is a solution in the case $F = 0$.
First we consider the case where $f = 0$. Then we have

$$u(x_0, y_0, z_0, t) = (t/4\pi) \int_\omega g(x_0 + \alpha_1 at, \, y_0 + \alpha_2 at, \, z_0 + \alpha_3 at) \, d\omega. \tag{11.5.1}$$

Since there is no difficulty differentiating twice under the integral sign, it is evident that $g \in C^2 \Rightarrow u \in C^2$. The first boundary condition

$$u(x_0, y_0, z_0, 0) = 0 \tag{11.5.2}$$

is quite evidently satisfied by (11.5.1). To verify the second boundary condition, differentiate the product in (11.5.1), noting that one (bounded) term

186

has a factor of t so that it vanishes at $t = 0$ while the other takes on the value

$$u_t(x_0, y_0, z_0, 0) = g(x_0, y_0, z_0) \tag{11.5.3}$$

at $t = 0$.

To verify that (11.5.1) satisfies the differential equation we use the second Green identity with $v = g$, $u = 1$ as follows, where $S = S(P, at)$:

$$u_t(x_0, y, z_0\, t) = \frac{1}{4\pi} \int_\omega g(x_0 + \alpha_1 t, \dots)\, d\omega + \frac{1}{4\pi at} \int_{\delta S} \frac{\partial g}{\partial n}(\cdots)(at)^2\, ds$$

$$= \frac{1}{4\pi} \int_\omega g(x_0 + \alpha_1 t, \dots)\, d\omega + \frac{1}{4\pi at} \int_{\delta S} \frac{\partial g}{\partial n}(\cdots)\, ds$$

$$= \frac{1}{4\pi} \int_\omega g(x_0 + \alpha_1 t, \dots)\, d\omega + \frac{1}{4\pi at} \int_S \Delta g\, dA$$

$$= \frac{1}{4\pi} \int_\omega g(x_0 + \alpha_1 t, \dots)\, d\omega + \frac{1}{4\pi at} \int_0^{at} r^2\, dr \int_\omega \Delta g\, d\omega.$$

Then

$$u_{tt} = -\frac{1}{4\pi a t^2} \int_S \Delta g\, dA + \frac{1}{4\pi} \int_\omega g_t(\cdots)\, d\omega + \frac{1}{4\pi t} \int_\omega \Delta g(\cdots) \cdot a^2 t^2\, d\omega$$

$$= a\left\{ -\frac{1}{4\pi a^2 t^2} \int_S \Delta g\, dA + \frac{1}{4\pi} \int_\omega g_n(\cdots)\, d\omega \right\} + \frac{a^2 t}{4\pi} \int_\omega \Delta g(\cdots)\, d\omega$$

$$= a^2 \Delta u.$$

Thus (11.5.1) does indeed satisfy the homogeneous differential equation. The quantity in braces is again zero by use of the second Green identity.

To show that our a priori representation is a solution when $F = 0$ and $g = 0$, but f possibly is not zero, we simply notice that this solution is obtained from (11.5.1) by replacing g with f and then taking $\partial/\partial t$ of the result. This is known as the Stokes rule (see Rayleigh's "Theory of Sound" Vol. I, p. 128; Vol. II, p. 99 [38]). Of course, because of the necessity to apply $\partial/\partial t$, we will need one more continuous derivative in f than in g. Thus we ask that $f \in C^3$.

6 VERIFICATION OF THE SOLUTION TO THE HOMOGENEOUS BOUNDARY-VALUE PROBLEM

The method of Stöss and Duhamel (see Courant and Hilbert [5, Vol. II]) utilizes an argument that is much like that used above but one that is much more sophisticated. It is, in a way, like the Stokes rule, but it is far less apparent; it states the solution of our problem where $f = g = 0$, but F possibly does not equal zero, can be obtained by replacing $g(x, y, z)$ in (11.5.1) by $a^2 F(x, y, z, t)$ with $t = \tau$ (fixed) and then integrating (rather than differentiating). More precisely, it says that with $f = g = 0$, (11.3.8) can be written as

$$W(x, y, z, t) = \int_0^t w(x, y, z, t; \tau) \, d\tau \qquad (11.6.1)$$

where for every $t > \tau$ and for every $(x, y, z) \in E^3$, $w(x, y, z, t; \tau)$ satisfies

$$w_{xx} + w_{yy} + w_{zz} - a^{-2} w_{tt} = 0, \qquad (11.6.2)$$

and for every $(x, y, z) \in E^3$

$$w(x, y, z, \tau; \tau) = 0, \qquad w_t(x, y, z, \tau; \tau) = -a^2 F(x, y, z, \tau). \qquad (11.6.3)$$

Of course, if w satisfies (11.6.2), (11.6.3), then from (11.5.1) and (11.6.1)

$$W(x, y, z, t) = -\frac{a^2}{4\pi} \int_0^t (t - \tau) \, d\tau \int_\omega F(x + \alpha_1 a(t - \tau), \, y + \alpha_2 a(t - \tau),$$

$$z + \alpha_3 a(t - \tau), \tau) \, d\omega.$$

Here putting $r = a(t - \tau)$, we have $\tau = t - r/a$ and

$$W(x, y, z, t) = -\frac{a^2}{4\pi} \int_{at}^0 \frac{r}{a} \left(-\frac{1}{a} \, dr \right) \int_\omega F\left(x + \alpha_1 r, y + \alpha_2 r, z + \alpha_3 r, t - \frac{r}{a} \right) d\omega$$

$$= -\frac{1}{4\pi} \int_0^{at} r^2 \, dr \int_\omega \frac{F(x + \alpha_1 r, y + \alpha_2 r, z + \alpha_3 r, t - r/a)}{r} \, d\omega$$

$$= -\frac{1}{4\pi} \int_{S(at, P)} \frac{F(Q, t - r/a)}{r} \, ds_Q$$

which verifies the contention that (11.6.1) is the solution (11.3.8) with $f = g = 0$.

We now verify that (11.6.1) satisfies the nonhomogeneous wave equation and homogeneous boundary conditions. We have from (11.6.1) and (11.6.3) that

$$W_t(x, y, z, t) = \int_0^t w_t(x, y, z, t; \tau) \, d\tau + w(x, y, z, t; t)$$

$$= \int_0^t w_t(x, y, z, t; \tau) \, d\tau. \qquad (11.6.4)$$

Then, again from (11.6.3)

$$W_{tt}(x,y,z,t) = \int_0^t w_{tt}(x,y,z,t;\tau)\, d\tau + w_t(x,y,z,t;t)$$

$$= \int_0^t w_{tt}(x,y,z,t;\tau)\, d\tau - a^2 F(x,y,z,t)$$

and

$$\Delta W(x,y,z,t) = \int_0^t \Delta w(x,y,z,t;\tau)\, d\tau.$$

In total, from (11.6.2)

$$\Delta W - \frac{1}{a^2}W_{tt} = \int_0^t \left(\Delta w - \frac{1}{a^2}w_{tt}\right) d\tau + F(x,y,z,t)$$

$$= F(x,y,z,t).$$

Also, from (11.6.1) and (11.6.4) it is obvious that

$$W_t(x,y,z,0) = W(x,y,z,0) = 0.$$

The Stöss–Duhamel procedure might be styled as a modified regular perturbation procedure and Stokes's rule.

At this point we are deeply impressed with the manipulative capacities of the classical giants. They lacked neither formal ability nor cleverness.

7 THE HADAMARD METHOD OF DESCENT

Having solved the Cauchy problem for the wave equation in four-dimensional space–time, we can specialize to obtain this solution in three- and in two-dimensional space–time; but this "method of descent" is not an entirely trivial process to carry out. Clearly, it is not necessary either to solve these problems in this way since all the above can be repeated with $\ln r$ replacing $1/r$. This will give the solution in three-dimensional space–time, and the two-dimensional space–time case is the D'Alembert solution given as introductory material in our first chapter. We undertake descent, of course, because we find it instructive and because in particular it helps us to understand why there should be such an astounding qualitative difference in four-dimensional space–time and three-dimensional space–time that the first has the magnificent Huyghens principle while the second does not. What seems magnificent to us is this difference between two and three space dimensions, not the Huyghens principle itself, and it is precisely this difference which descent elucidates.

Pictures are easier to understand in E^3 to E^1 than in E^3 to E^2, so we first treat descent to two-dimensional space–time, although it will in itself be of rather little interest. For this purpose define functions f^*, g^*, F^* of one variable such that for every x, y, z,

$$f^*(x) = f(x, y, z), \qquad g^*(x) = g(x, y, z), \qquad F^*(x) = F(x, y, z);$$

i.e., let $f_y = f_z = g_y = g_z = F_y = F_z = 0$ and represent f, g, F as functions of one variable by using an asterisk superscript. We solve the corresponding Cauchy problem with

$$u(x, t) = U(x, 0, 0, t)$$

using descent on (11.3.8). Where $S = S(at, (x, 0, 0))$, let us evaluate

$$(1/4\pi a^2 t) \int_{\delta S} g(\xi, \eta, \zeta) \, dS, \tag{11.7.1}$$

remembering that g is independent of η and ζ (see Fig. 51). Any circular disk parallel to the yz plane of differential width ds has $g^*(\xi) = g(\xi, \eta, \zeta)$ over the entire width of the outer surface of the disk. Therefore,

$$g(\xi, \eta, \zeta) \, dS = g^*(\xi) \, 2\pi [a^2 t^2 - (\xi - x)^2]^{\frac{1}{2}} \, ds$$

where ds is an element of arclength and dS an element of surface area. Since the whole sphere is covered by integrating disks from $s = 0$ to $s = \pi a t$,

$$\frac{1}{4\pi a^2 t} \int_{\delta S} g(\xi, \eta, \zeta) \, dS = \frac{1}{4\pi a^2 t} \int_0^{\pi a t} g^*(\xi) [a^2 t^2 - (\xi - x)^2]^{\frac{1}{2}} \, 2\pi \, ds.$$

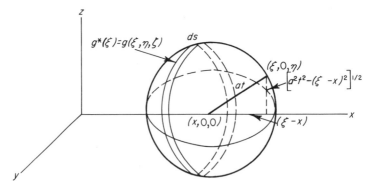

FIG. 51 $\bar{g}(x)$ is integrated over a sphere of radius at at the point $(x, 0, 0)$ to obtain the D'Alembert solution.

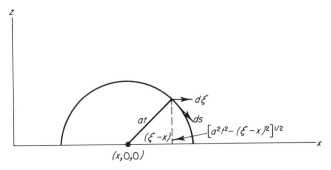

FIG. 52   Conversion from arclength to x or ξ integration.

We put the integral in terms of $d\xi$ (see Fig. 52). The indicated angles are complements of the same angles, so their cosines

$$d\xi/ds \qquad \text{and} \qquad [a^2 t^2 - (\xi - x)^2]^{1/2}/at$$

are equal. Thus

$$ds/d\xi = at/[a^2 t^2 - (\xi - x)^2]^{1/2},$$

and the first term of the solution is

$$\frac{1}{4\pi a^2 t} \int_{x-at}^{x+at} g^*(\xi) 2\pi at \, d\xi = \frac{1}{2a} \int_{x-at}^{x+at} g^*(\xi) \, d\xi.$$

Applying this analysis to all three terms in (11.3.8), we have

$$u(x,t) = \frac{1}{2a} \int_{x-at}^{x+at} g^*(\xi) \, d\xi + \frac{\partial}{\partial t} \frac{1}{2a} \int_{x-at}^{x+at} f^*(\xi) \, d\xi$$

$$- \frac{a}{2} \int_0^t d\tau \int_{x-a(t-\tau)}^{x+a(t-\tau)} F^*(\xi, \tau) \, d\xi. \tag{11.7.2}$$

For the homogeneous differential equation

$$u(x,t) = (1/2a) \int_{x-at}^{x+at} g^*(\xi) \, d\xi + \tfrac{1}{2}[f^*(x+at) + f^*(x-at)], \tag{11.7.3}$$

which is the familiar formula of D'Alembert (1.12).

In three-dimensional space–time, we have (see Fig. 53) that

$$\frac{d\xi \, d\eta}{dS} = \frac{[a^2 t^2 - (\xi - x)^2 - (\eta - y)^2]^{1/2}}{at}$$

191

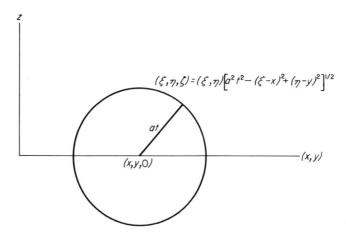

FIG. 53 Variables for descent to three-dimensional space–time.

or

$$ds = \frac{at \, d\xi \, d\eta}{[a^2 t^2 - (\xi - x)^2 - (\eta - y)^2]^{1/2}}.$$

Therefore, where $S = S(at, (x, y, 0))$ and $C = C(at, (x, y))$,

$$\frac{1}{4\pi a^2 t} \int_{\delta S} g(\xi, \eta, \zeta) \, dS = \frac{1}{4\pi a^2 t} \int_C \left(2 \frac{g^*(\xi, \eta) \, at}{[a^2 t^2 - (\xi - x)^2 - (\eta - y)^2]^{1/2}} \right) d\xi \, d\eta$$

$$= \frac{1}{2\pi a} \int_C \frac{g^*(\xi, \eta)}{[a^2 t^2 - (\xi - x)^2 - (\eta - y)^2]^{1/2}} \, d\xi \, d\eta.$$

All three terms of (11.3.8) give

$$u(x, y, t) = \frac{1}{2\pi a} \int_C \frac{g^*(\xi, \eta)}{[a^2 t^2 - (\xi - x)^2 - (\eta - y)^2]^{1/2}} \, d\xi \, d\eta$$

$$+ \frac{\partial}{\partial t} \left(\frac{1}{2\pi a} \int_C \frac{f^*(\xi, \eta)}{[a^2 t^2 - (\xi - x)^2 + (\eta - y)^2]^{1/2}} \, d\xi \, d\eta \right)$$

$$- \frac{a}{2\pi} \int_S \frac{F^*(\xi, \eta, \tau)}{[a^2 (t - \tau)^2 - (\xi - x)^2 - (\eta - y)^2]^{1/2}} \, d\tau. \quad (11.7.4)$$

EXERCISE 1 Using $v = \varphi$ [see (11.2.2)] and $u = \ln r_{PQ}$ in the second Green's identity, develop the Kirchhoff identity for $n = 2$. From this, develop directly the solution (11.7.4) for the Cauchy problem for the nonhomogeneous wave equation in three-dimensional space–time.

8 THE HUYGHENS PRINCIPLE

Keeping these matters in mind, we now discuss the properties of the solution of the Cauchy problem for the wave equation in four-dimensional space–time. For this purpose, we first examine the solution (11.4.3) expressed in mean-value form.

We have found that if Cauchy data is given in the region $A \subset E^3, f \in C^3(A)$, $g \in C^2(A)$, the solution of the homogeneous wave equation at any later time t is given in terms of integrals and time derivatives of them over the surface of a sphere of radius at (see Fig. 54), provided, of course, that this surface lies within A. This is the Huyghens principle; or rather, since the Cauchy problem for the wave equation in four-dimensional space–time has this property, it is said that it is of Huyghens' type. To say that the solution does not depend on values outside a sphere of radius at is only to say that there is a propagation rate a involved—which will surely surprise no one—but the fact that *the solution does not depend on values at interior points of the sphere $S(at, P)$* is a nontrivial observation.

Before we consider this phenomenon in other dimensions, however, we undertake some clarifications of it in four-dimensional space–time. First, we note that the same facts hold true for the nonhomogeneous wave equation.

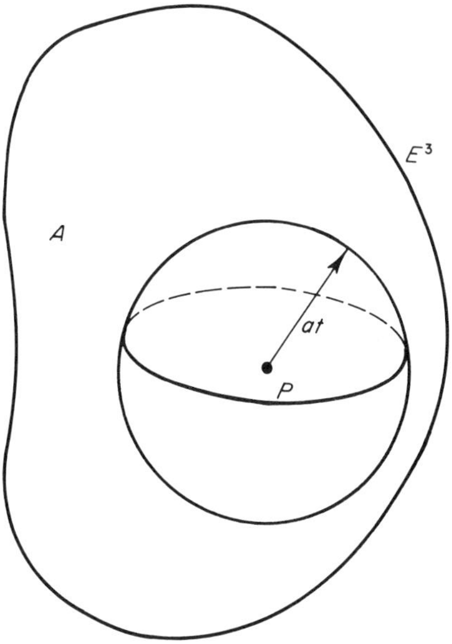

FIG. 54   The solution at a point $P \in A$ depends only on values on the surface of $S(at, P)$.

We have seen that if $F \neq 0$, the term to be addended is

$$-\frac{1}{4\pi} \int_S \frac{F(Q, t - r/a)}{r} \, dA, \tag{11.8.1}$$

where $S = S(at, P)$. This is, of course, a volume, not a surface, integral. But notice should be taken that the integration is over a sphere given by

$$\{(x, y, z) \mid (x-y_0)^2 + (y-y_0)^2 + (z-z_0)^2 \leqslant (at)^2, \quad P = (x_0, y_0, z_0)\}$$

while F is a function of t as well as it is a function of (x, y, z); in the integrand of (11.8.1) only values of F at $t^* = t - r/a$ appear, and this has the effect of giving us values only over a shell of radius $r = at$. The nonhomogeneous case will not again be mentioned here and is now discussed only for its passing interest.

Let us reverse our view and ask about where the effect is felt, with passage of time, of the initial data given in A, rather than asking what data affects the solution at a particular point. For clarity in this, consider a disturbance on a "spot" in A (see Fig. 55), which may be made as small as desired and which may, therefore, be thought of as a point disturbance. It is still required that $f \in C^3(A)$ and $g \in C^2(A)$, so that the theory developed in this chapter will apply, but it is asked that f and g together with their derivatives vanish at the boundary and exterior points of the spot. From the above, at time t only points a distance at from the points of the spot will be affected by the "spot disturbance," and only in a shell with thickness equal to the diameter of the spot will there be any nonzero solution. Letting this diameter, and therefore, thickness, shrink to zero, we see that a spherical wave front is traveling through the medium of A, and this is the phenomenon described in physics texts as the Huyghens principle.

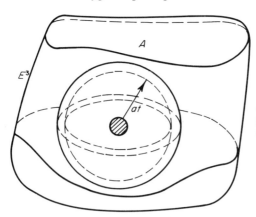

FIG. 55 Data are given in A which is "sufficiently smooth" there, but nonzero only inside a small "spot."

Once again, let it be noted that the remarkable fact is not that the solution is zero in advance of the wave, since one expects a true propagation phenomena, but that it is zero behind the advancing wave. To explain the origin of an initial disturbance in a spot, one may think of it as arising from a forcing function F which is nonzero on the spot until time $t = 0$ when it becomes zero and remains so, leaving the homogeneous wave equation to be solved for the disturbing effect of the time when F was not zero on the spot. The solution would suddenly start to be zero inside the sphere of radius at. It is this sharp "cut-off propagation" that typifies the Huyghens principle, rather than the "sharp turn-on" which typifies true propagation phenomena in general and which holds for dimensions where the Huyghens principle is not valid.

It became quite clear [see (11.7.3), (11.7.4)] in the last section that the Huyghens principle does not operate in a two- or three-dimensional space–time. Obtaining these results by Hadamard's beautiful "descent" exposition allows us to see how this can be so, even though it does operate in four-dimensional space–time. The descent, for example, from four- to three-dimensional space–time clearly shows the integrals over the *surface of the sphere* $S(at,(x,y,z))$ in E^3 being smashed down, as between two blocks of wood, to integrals over the *interior of the circle* $C(at,(x,y))$ in E^2. It will, however, be instructive to examine the nature of the, so to speak, residual solution left behind an advancing wave in two- and three-dimensional space–time. It will be found that while there is a permanent set left by an advancing wave in two-dimensional space–time, there is a residue left in three-dimensional space–time which decays with time. Thus three-dimensional space–time will be seen to be superior to two-dimensional space–time in respect to this property. In this respect, two-dimensional space–time is a very special case, not suited for demonstration of the Huyghens principle at all, except in the case that the initial rate of displacement is given to be zero.

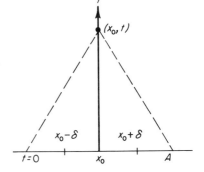

FIG. 56 A permanent set is left in an infinite string by a disturbance on a spot.

Let us look at the effect of a spot disturbance on an infinite taut string. We ask that in this spot, $(x_0 - \delta, x + \delta)$, f and g be positive, but that $f = g = 0$ at all other points of the string. We observe the displacement record at a point x_0 inside the spot as time passes. Of course, at first the D'Alembert solution gives a contribution from both f and g (see Fig. 1), but as time passes [in Fig. 56, as (x_0, t) moves vertically], the characteristics from (x_0, t) eventually cut back across the line $t = 0$ at points outside the interval of disturbance and thus at points where $f = 0$. The D'Alembert solution (11.7.3) or (1.1.12) then shows there is no further contribution to the displacement u from f. Call the time t^* when the characteristics from (x_0, t) intersect the line $t = 0$ at $x = x_0 - \delta$ and $x = x_0 + \delta$. Then, for all $t > t^*$,

$$u(x_0, t) = \tfrac{1}{2} \int_{x_0 - \delta}^{x_0 + \delta} g(s)\, ds, \tag{11.8.2}$$

a constant not depending on t, and, because we required that g be positive in $(x_0 - \delta, x_0 + \delta)$, a positive constant. This is the permanent set at x_0 we have spoken of above. It should be quickly added that this could only be physically valid for an infinite string. If the string is attached to a fixed point, say, the reflected wave will eventually return and wipe out the "permanent" displacement at x_0. Nevertheless, it would remain until a reflected wave came to annihilate it.

Now let us think of a taut membrane, say a drumhead. In an infinite membrane, a spot disturbance would leave forever at the point (x_0, y_0) the displacement

$$u(x_0, y_0, t) = \frac{1}{2\pi a} \int_{C*} \frac{g(\xi, \eta)}{[a^2 t^2 - (\xi - x_0)^2 - (\eta - y_0)^2]^{\frac{1}{2}}}\, d\xi\, d\eta$$
$$+ \frac{\partial}{\partial t} \left(\frac{1}{2\pi a} \int_{C*} \frac{f(\xi, \eta)}{[a^2 t^2 - (\xi - x_0)^2 - (\eta - y_0)^2]^{\frac{1}{2}}}\, d\xi\, d\eta \right), \tag{11.8.3}$$

where $C* = C(at^*, (x_0, y_0))$, and t^* is the time at which the characteristic cone from the point (x_0, y_0, t_0),

$$\{(x, y, t) \mid a^2 (t - t_0)^2 = (x - x_0)^2 + (y - y_0)^2\},$$

intersects the surface $t = 0$ in a circle which contains the initial spot disturbance. *However, the displacement* (11.8.3), *unlike that in* (11.8.2), *depends still on time t, and, moreover, it is seen then that* (11.8.3) *does asymptotically return to its original equilibrium value.* This is typical of behavior in odd-dimensional space–time.

A principle of great importance in mathematical physics is that, with the exception of two-dimensional space–time which has its own peculiar behavior, even-dimensional space–time always exhibits a Huyghens principle while odd-dimensional space–time does not. Recent investigations have been directed toward ferreting out those principles which cause four-dimensional space–time to play a special role in our universe that is not shared with space–time in other even space–time dimensions. To further one's understanding of the point about even- and odd-dimensional space–time with respect to the Huyghens principle we suggest the following list of exercises.

EXERCISE 2 Using $v = \varphi$ [see (11.2.2)] and $u = (1/r_{PQ})^{n-1}$ in the second Green identity, develop the Kirchhoff identity for $n \geqslant 3$ just as it was done in Section 2 for $n = 3$ and just as Exercise 1 in Section 7 asks for it to be done for $n = 2$.

EXERCISE 3 Develop the solution of the Cauchy problem for the homogeneous wave equation for $n \geqslant 3$ as it was done in Section 3 for $n = 3$.

EXERCISE 4 A spherically symmetric solution of the wave equation,

$$u_t = \sum_{i=1}^{n} u_{x_i x_i},$$

is a solution that depends only on t and $\rho = (\sum_{i-1}^{n} x_i^2)^{1/2}$. Put this equation in spherically symmetric form; i.e., write the wave equation as an equation in two independent variables t and ρ, eliminating all terms containing derivatives of u with respect to the angular variables in spherical coordinates.

EXERCISE 5 Using the solutions already derived for the wave equation for $n \geqslant 2$, write solutions for the spherically symmetric wave equation for $n \geqslant 2$.

EXERCISE 6 Show that the Huyghens principle is valid for $n+1$ even.

EXERCISE 7 Find the nature of the residual decay for $n+1$ odd, $n \geqslant 2$. Is it of the same nature or faster or slower than for $n = 2$?

IV

Boundary-Value Problems for Equations of Elliptic–Parabolic Type

The class of boundary-value problems (5.3.8) presented in Chapter 5, Section 3, forms the context of all that follows. This theory is the creation of Fichera [13]. We will provide here his proof for the existence of $\mathscr{L}^p$-weak solutions (see Chapter 5, Section 8) using the abstract existence principle, which is a theorem in functional analysis also due to Fichera [14] and designed to be used for such purposes. Since we aim at completeness to some extent, all that is needed in functional analysis will be presented here in brief. It will be found that the abstract existence principle establishes $\mathscr{L}^p$-weak existence immediately, provided only that certain inequalities for regular functions are proven to be valid. Of course, to be useful such inequalities must be proven without recourse to any assumption of existence of solutions to our boundary-value problems, and they are therefore called a priori inequalities. It will be noted that in itself the abstract existence principle is altogether independent of type, so that the role of type, characteristics, style of boundary-data specification, and such things are all to be reflected in the nature of the a priori inequalities for linear equations.

The basic tool used to obtain almost any such inequalities is simply the generalized Green identity (5.7.4) again, but the details of techniques used to obtain these results are one of the most intriguing parts of the work. In broad outline, however, we proceed much as we did in Part III to obtain a priori inequalities[†] for the Laplace, Poisson, and nonhomogeneous wave equations, some of which, in fact, turned out in Chapter 11 to be solutions. The a priori "estimates" which are to be developed here, however, will not be found to be so very "close to the truth." We will derive a remarkable a priori maximum principle and other features similarly useful in the computation of bounds for errors of approximation (see Chapter 3, Section 5). The same inequalities are useful, of course, in proving the uniqueness of regular solutions for problems of elliptic–parabolic type; the uniqueness of $\mathscr{L}^p$-weak solutions is the work of Olga Oleinik, a Russian analyst who models extremely clever constructions very closely to those of Fichera. We will first consider derivations of the basic inequalities needed, but some preliminaries must be presented.

[†] Equalities are thought of as special cases of inequalities when they are not taken in a strict sense.

12

A Priori Inequalities

1 SOME PRELIMINARIES

The $\lim_{p \to \infty} a^{1/p}$ plays a crucial role in what follows.

Proposition 1 $a > 0$ implies $\lim_{p \to \infty} a^{1/p} = 1$.

Proof

$$\lim_{p \to \infty} a^{1/p} = \lim_{p \to \infty} e^{(1/p)\ln a} = e^0 = 1.$$

Proposition 2 $f \in C[a, b]$ implies

$$\lim_{p \to \infty} \left(\int_a^b |f(x)|^p \, dx \right)^{1/p} = \max_{a,b} |f(x)|.$$

Proof (a) Let $M = \max_{[a,b]} |f(x)|$. Then

$$\left(\int_a^b |f(x)|^p \, dx \right)^{1/p} \leqslant (M^p(b-a))^{1/p} = M(b-a)^{1/p},$$

and, from Proposition 1, the limit is less than or equal to M.

(b) There exists $x^* \in [a, b]$ such that $M = f(x^*)$ (see Fig. 57). Then, from

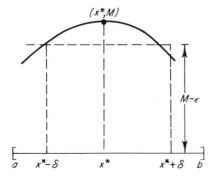

FIG. 57 A 2δ neighborhood about the point of maximum of f.

the continuity of f, for every $\varepsilon > 0$, there exists $\delta > 0$ such that

$$\left(\int_a^b |f(x)|^p \, dx\right)^{1/p} \geqslant \left(\int_{x^*-\delta}^{x^*+\delta} |f(x)|^p \, dx\right)^{1/p}$$
$$\geqslant ((M-\varepsilon)^p (2\delta))^{1/p} = (M-\varepsilon)(2\delta)^{1/p}.$$

Therefore, by Proposition 1,

$$\lim_{p \to \infty} \left(\int_a^b |f(x)|^p \, dx\right)^{1/p} \geqslant M - \varepsilon.$$

If $x^* = a$ or $x^* = b$, then one limit of the integral, either $x^* - \delta$ or $x^* + \delta$, can be replaced by a or b, and 2δ replaced by δ. Of course, since ε is arbitrary, the proposition is established.

Let R_+^1 be the set of positive real numbers and let $L^1(a,b)$ be the set of functions whose absolute value is integrable.

Proposition 3 $f \in C[a,b]$ and $W: (a,b) \to R_+^1$ and $W \in L^1(a,b)$ implies

$$\lim_{p \to \infty} \left(\int_a^b |f(x)|^p W \, dx\right)^{1/p} = \max_{a,b} |f(x)|$$

Proof The proof is identical to Proposition 2 except that $\int_a^b w \, dx$ now plays the role of $b-a$. One should notice that the condition $w \geqslant 0$ is essential in this proof.

Proposition 4 $f \in C(a,b)$ and f bounded on (a,b) and $W: (a,b) \to R_+^1$ and $w \in L^1(a,b)$ implies

$$\lim_{p \to \infty} \left(\int_a^b |f(x)|^p w \, dx\right)^{1/p} = \operatorname*{lub}_{(a,b)} |f(x)|.$$

Proof Since a bounded set (the range of f) of real numbers has a least upper bound, the proof of Proposition 2 will do here with lub replacing max.

Cauchy–Riesz Lemma A is a region[†] of positive measure[‡] and $W: A \to R_+{}^1$, $W \in L^1(A)$ and f is bounded in A implies

$$\lim_{p \to \infty} \left(\int_A |f|^p W \, dA \right)^{1/p} = \lub_A |f|.$$

The proof is a direct adaptation of the above.

Schwarz–Hölder Inequality For $p, q \geqslant 1$ and $1/p + 1/q = 1$ and $u \in L^p(A)$, $v \in L^q(A)$,

$$\int_A uv \, dA \leqslant \left(\int_A |u|^p \, dA \right)^{1/p} \left(\int_A |v|^q \, dA \right)^{1/q}.$$

EXERCISE 1 Prove the above. The Schwarz inequality of Chapter 10, Section 9, for quadratic functionals gives this inequality for $p = q = 2$.

2 A PROPERTY OF SEMIDEFINITE QUADRATIC FORMS

From review of Chapter 5, Section 3, the student will recall that an operator

$$Lu = a^{ij} u_{x_i x_j} + b^i u_{x_i} + cu,$$

defined on A, is said to be of elliptic–parabolic type on A if

$$\text{for every} \quad \lambda_i, \lambda_j \in R^1, \qquad a^{ij} \lambda_i \lambda_j \geqslant 0,$$

and that on those portions of the boundary δA of A designated by $\Sigma^1 \cup \Sigma^2$ (the characteristic portions), $a^{ij} n_i n_j = 0$. We will find the following lemma useful in considering the values of boundary terms after application of the Green identity to operators of this type.

Let n_i be the direction cosines of the inner normal to the boundary Σ of a region A at the point $x \in \Sigma$.

Lemma For every $i, j = 1, \ldots, n$, $(a^{ij}(x) = a^{ji}(x))$ and $(a^{ij}(x) \lambda_i \lambda_j \geqslant 0$ for every $\lambda_i, \lambda_j \in R^1)$ and $(a^{ij}(x) n_i n_j = 0)$ implies $a^{ij}(x) n_i = 0$.

---

† Here always an open connected set.

‡ One may think of this as area or volume or simply as $\int_A 1 \, dA$ if he has not already studied integration theory.

Proof Let $A = (a^{ij})$ be a real symmetric matrix. As such, it represents a linear transformation on R^n into R^n. Adopting the unit eigenvectors $V_1, V_2, \ldots, V_n$ of A as a set of basis elements in R^n, this linear transformation can be represented by a diagonal matrix B (said to be "similar" to A),

$$B = \begin{bmatrix} v_1 & 0 & \cdots & 0 \\ 0 & v_2 & \cdots & 0 \\ \vdots & & \ddots & \vdots \\ 0 & \cdots & 0 & v_n \end{bmatrix}$$

where the diagonal elements v_i are eigenvalues of A corresponding to V_i. For every $y \in R^n$, we must have

$$y^T B y \geqslant 0;$$

thus if y_i is a vector with ith component equal to one and all others zero,

$$v_i = y_i^T B y_i \geqslant 0. \tag{12.2.1}$$

Since the V_i span R^n, the normal $\bar{n}(x)$ to Σ at a point $x \in \Sigma$ (where it exists) can be written

$$\bar{n}(x) = a_1 V_1 + \cdots + a_n V_n,$$

and we have

$$B\bar{n}(x) = a_1 v_1 V_1 + \cdots + a_n v_n V_n. \tag{12.2.2}$$

By hypothesis,

$$(\bar{n}(x), B\bar{n}(x)) = 0 = a_1^2 v_1 + \cdots + a_n^2 v_n. \tag{12.2.3}$$

We wish to show that (12.2.2) is zero. By (12.2.1), all the nonzero terms in (12.2.3) contain factors $v_i > 0$. Then, since $a_i^2 \geqslant 0$, the equality (12.2.3) implies that $a_i = 0$ for all i such that $v_i \neq 0$. Thus all terms in (12.2.2) are zero either because they contain a factor $v_i = 0$ or a factor $a_i = 0$.

3 THE GENERALIZED GREEN IDENTITY USING $v = (u^2 + \delta)^{p/2}$

Let A be a Green region, $A \subset E^n$, with topological boundary Σ having an *inner normal* $n(x) = (n_i)$ (see Fig. 58) defined at all points $x \in \Sigma$ except possibly on a set of measure zero. With

$$Lu = a^{ij}(x) u_{x_i x_j} + b^i(x) u_{x_i} + c(x) u \quad \text{and}$$

$$L^* u = (a^{ij}(x) u)_{x_i x_j} - (b^i(x) u)_{x_i} + c(x) u,$$

FIG. 58   For consistency with Fichera's notation we use an inner normal.

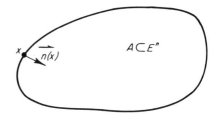

we have from (5.7.4), (5.7.5) that for every v, $w \in C^2(A)$, $C^1(A \cup \Sigma)$,

$$\int_A (vL^*w - wLv)\, dA = \int_\Sigma [w(\partial v/\partial v + bv) - v\, \partial w/\partial v]\, ds, \qquad (12.3.1)$$

where for every $x \in \Sigma$,

$$b(x) = (b^i(x) - a^{ij}_{x_j}(x)n_i), \qquad (12.3.2)$$

and, letting u be v or w,

$$\partial u(x)/\partial v = a^{ij}(x)u_{x_j}n_i. \qquad (12.3.3)$$

Of course, we can write

$$\partial u(x)/\partial v = (\operatorname{grad} u, v) \qquad (12.3.4a)$$

where

$$v(x) = (v_j) = (a^{ij}(x)n_i). \qquad (12.3.4b)$$

Such a vector is called the conormal, and it can be clearly seen in (12.3.1) that, with respect to the Green identity, the conormal derivative plays a role for all linear second-order partial differential operators, which replaces that of the normal derivative for the Laplace operator. When the lower order coefficients b^i are zero and the a^{ij} are constant, then the role is identical. A basic point in the treatment of elliptic–parabolic type is that from the lemma in Section 2, for such operators, if x is a point of the characteristic portion $\Sigma^1 \cup \Sigma^2$ of the boundary Σ, then

$$v(x) = 0, \qquad \text{so that} \quad \partial u(x)/\partial v = 0. \qquad (12.3.5)$$

Here, of course, u stands for v or w.

We now let

$$v = (u^2 + \delta)^{p/2}, \qquad (12.3.6)$$

where $u \in C^2(A)$, $\delta > 0$, and $p \geqslant 1$, and examine the left member of the generalized Green identity (12.3.1). We have

205

$$Lv = a^{ij}((u^2+\delta)^{p/2})_{x_ix_j} + b^i((u^2+\delta)^{p/2})_{x_i} + c(u^2+\delta)^{p/2}$$

$$= a^{ij}[(p/2)(2uu_{x_i})(u^2+\delta)^{(p/2)-1}]_{xj} + (p/2)b^i(2uu_{x_i})(u^2+\delta)^{(p/2)-1}$$
$$+ c(u^2+\delta)^{p/2}$$

$$= a^{ij}[(p/2)(2uu_{x_i})(u^2+\delta)^{(p/2)-1}]_{x_j}$$
$$+ (p/2)b^i(2uu_{x_i})(u^2+\delta)^{(p/2)-1} + c(u^2+\delta)^{p/2}$$

$$= pa^{ij}[u_{x_j}u_{x_i}(u^2+\delta)^{(p/2)-1} + u \cdot u_{x_ix_j}(u^2+\delta)^{(p/2)-1}$$
$$+ uu_{x_i}(p-2)uu_{x_j}(u^2+\delta)^{(p/2)-2}]$$
$$+ pb^i uu_{x_i}(u^2+\delta)^{(p/2)-1} + c(u^2+\delta)^{p/2}$$

$$= p[(p-2)u^2 + u^2 + \delta](u^2+\delta)^{(p/2)-2}a^{ij}u_{x_i}u_{x_j}$$
$$+ up(u^2+\delta)^{(p/2)-1}a^{ij}u_{x_ix_j} + up(u^2+\delta)^{(p/2)-1}b^i u_{x_i}$$
$$+ up(u^2+\delta)^{(p/2)-1}cu - up(u^2+\delta)^{(p/2)-1}cu$$
$$+ c(u^2+\delta)(u^2+\delta)^{(p/2)-1}$$

$$= p[(p-1)u^2 + \delta](u^2+\delta)^{(p/2)-2}a^{ij}u_{x_i}u_{x_j} + p(u^2+\delta)^{(p/2)-1}uLu$$
$$+ c(u^2+\delta)^{(p/2)-1}[(1-p)u^2 + \delta].$$

Therefore,

$$\int_A (vL^*w - wLv)\, dA = \int_A [(u^2+\delta)^{p/2}L^*w$$
$$- w\{p[(p-1)u^2+\delta](u^2+\delta)^{(p/2)-2}a^{ij}u_{x_i}u_{x_j}$$
$$+ p(u^2+\delta)^{(p/2)-1}uLu$$
$$+ c(u^2+\delta)^{(p/2)-1}[(1-p)u^2 + \delta]\}]\, dA.$$

We now examine the right member of (12.3.1) using the result (12.3.5):

$$\int_\Sigma (w[\partial v/\partial v + bv] - v\,\partial w/\partial v)\, ds = \int_{\Sigma^1 \cup \Sigma^2} \{w[\partial v/\partial v + bv] - v\,\partial w/\partial v\}\, ds$$
$$+ \int_{\Sigma^3} \{w[\partial v/\partial v + bv] - v\,\partial w/\partial v\}\, ds$$
$$= \int_{\Sigma^1 \cup \Sigma^2} (bvw)\, ds$$
$$+ \int_{\Sigma^3} \{w[\partial v/\partial v + bv] - v\,\partial w/\partial v\}\, ds$$

Putting $v = (u^2+\delta)^{p/2}$, we have from (12.3.1)

$$\int_A \{(u^2+\delta)^{p/2} L^*w - w[p[(p-1)u^2+\delta](u^2+\delta)^{(p/2)-2} a^{ij} u_{x_i} u_{x_j}$$

$$+ p(u^2+\delta)^{(p/2)-1} uLu + c(u^2+\delta)^{(p/2)-1} [(1-p)u^2+\delta]]\} \, dA$$

$$= \int_{\Sigma^1 \cup \Sigma^2} w(u^2+\delta)^{p/2} b \, ds + \int_{\Sigma^3} [p(u^2+\delta)^{(p/2)-1} wua^{ij} u_{x_i} n_j$$

$$- (u^2+\delta)^{p/2} (a^{ij} w_{x_i} n_j - bw)] \, ds. \tag{12.3.7}$$

This is the (corrected) equation (3.3) appearing in Fichera's work [13].

4 A FIRST MAXIMUM PRINCIPLE

We will develop a sequence of lemmas that yields a maximum principle for the homogeneous boundary-value problem for the nonhomogeneous equation. Let L be of elliptic–parabolic type.

L-regular If $u \in C^2(A)$, $C^1(A \cup \Sigma)$ and Lu is bounded in A, then we say that u is L-regular in A and write $u \in C_L(A)$.

Lemma 1 (i) $p \in [1, \infty) \wedge$ (ii) $w \in C^2(A \cup \Sigma) \ni W \leqslant 0 \wedge L^*W + (p-1)cw > 0$ in $A \cup \Sigma \wedge$ (iii) $u \in C_L(A) \wedge$ (iv) $u = 0$ a.e. in $\Sigma^2 \cup \Sigma^3 \Rightarrow$

$$\left(\int_A |u|^p \, dA \right)^{1/p} \leqslant p \frac{\max_{A \cup \Sigma} |w|}{\min_{A \cup \Sigma} [L^*w + (p-1) cw]} \left(\int_A |Lu|^p \, dA \right)^{1/p}. \tag{12.4.1}$$

Proof In (12.3.7), let $=$ be replaced by $\leqslant$; then any quantities in the left member which are nonnegative may be dropped and any quantities in the right member that are nonpositive may be dropped.

In the left member $a^{ij} u_{x_i} u_{x_j} \leqslant 0$, because L is of elliptic–parabolic type, and by hypothesis $w \leqslant 0$ so the product of these two functions may be dropped.

In the right member the integral over Σ^1 may be dropped because on Σ^1 we have $b > 0$ [see (5.3.7)], and since $w \leqslant 0$, the product is nonpositive. Also, according to hypothesis (iv), u may be dropped from integrals over Σ^2 or Σ^3. These deletions leave the equation

$$\int_A (u^2+\delta)^{p/2} \{L^*w - cw(u^2+\delta)^{-1}[(1-p)u^2+\delta]\} \, dA$$

$$\leqslant p \int_A (u^2+\delta)^{(p/2)-1} uwLu \, dA$$

$$+ \delta^{p/2} \left[\int_{\Sigma^2} wb \, ds - \int_{\Sigma^3} (a^{ij} w_{x_i} n_j - bw) \, ds \right]. \tag{12.4.2a}$$

207

We observe that $\lim_{\delta \to 0}$ of the integrand of the left member is

$$(u^2)^{p/2}(L^*w - (1-p)\,cw) = |u|^p(L^*w - (1-p)\,cw),$$

while the factor $(u^2 + \delta)^{(p/2)-1}u$ in the right member approaches

$$(u^2)^{(p/2)-1}u = |u|^{p-1}$$

uniformly with respect to x in $A \cup \Sigma$.

Since the boundary terms vanish with δ, taking absolute values of both sides and using the triangle inequality,

$$\int_A |u|^p \, dA \leqslant p \frac{\max_{A \cup \Sigma}|w|}{\min_{A \cup \Sigma}[L^*(w) + (p-1)\,cw]} \int_A |u|^{p-1}\,|Lu| \, dA, \quad (12.4.2b)$$

which for $p = 1$ is the same as the lemma. To obtain it for $p > 1$, we use the Schwarz–Hölder inequality with $q = p/(p-1)$ on the right member of (12.4.2):

$$\int_A |u|^{p-1}\,|Lu| \, dA \leqslant \left(\int_A |Lu|^p \, dA\right)^{1/p}\left(\int_A (|u|^{p-1})^{p/(p-1)} \, dA\right)^{(p-1)/p}$$

$$= \left(\int_A |Lu|^p \, dA\right)^{1/p} \cdot \left(\int_A |u|^p \, dA\right)^{(p-1)/p}.$$

Then from (12.4.2b)

$$\int_A |u|^p \, dA \leqslant p \frac{\max_{A \cup \Sigma}|w|}{\min_{A \cup \Sigma}[L^*w + (p-1)\,cw]}\left(\int_A |Lu|^p \, dA\right)^{1/p}\left(\int_A |u|^p \, dA\right)^{(p-1)/p}$$

or, dividing both sides by $(\int_A |u|^p \, dA)^{(p-1)/p}$, the lemma is established.

Remark The inequality (12.4.1) is an a priori inequality in L^p-norm for $u = 0$ a.e. on $\Sigma^2 \cup \Sigma^3$, whenever an appropriate function w can be found. Assuming only that $c < 0$ (as we have already done in Chapter 5, Section 3), such functions can be difficult to find, though it is evident that they do exist. In some cases, however, the trivial choice $w = -1$ will work:

(i) $c < 0$ and p is large enough that

$$\min_{A \cup \Sigma} [-a^{ij}_{x_i x_j} + b^i_{xi} - pc] < 0,$$

(ii) $c^* < 0$ and p is small enough that

$$\min_{A \cup \Sigma} [(1-p)c - c^*] > 0,$$

(iii) $c < 0$ and $c^* < 0$ and $p \geqslant 1$.

The L^p a priori inequality (12.4.1) will be used to prove our first maximum principle by letting $p \to \infty$, but we need to rework it slightly.

Let $R_-{}^1$ be the set of negative real numbers.

Lemma 2 $(c, w: A \cup \Sigma \to R_-{}^1) \wedge (w \in C^2(A \cup \Sigma)) \wedge (\exists p_0 \in R^1 \ni L^*w + (p_0 - 1)cw \geqslant 0$ in $A \cup \Sigma) \wedge (p > 1) \wedge (p > p_0) \wedge (u \in C_L) \wedge (u = 0$ a.e. on $\Sigma^2 \cup \Sigma^3) \Rightarrow$

$$\left(\int_A |u|^p cw \, dA \right)^{1/p} \leqslant \frac{p}{p - p_0} \left(\int_A \left| \frac{L(u)}{c} \right|^p cw \, dA \right)^{1/p}. \qquad (12.4.3)$$

Proof We have that

$$(cw > 0) \wedge (p > p_0) \wedge L^*w + (p-1)cw > L^*w + (p_0 - 1)cw \geqslant 0,$$

so the hypotheses of Lemma 1 are satisfied. From (12.4.2a), letting $\delta \to 0$, we have

$$\int_A (u^2)^{p/2} [L^*w - (1-p)cw] \, dx \leqslant p \int_A (u^2)^{(p/2)-1} uwLu \, dA.$$

Then from

$$L^*w - (1-p)cw = [L^*w + (p_0 - 1)cw] + (p - p_0)cw$$
$$\geqslant (p - p_0)cw,$$

it follows that

$$\int_A |u|^p cw \, dA \leqslant \frac{p}{p - p_0} \int_A |u|^{p-1} |w| |Lu| \, dA$$

$$= \int_A |u|^{p-1} |c|^{(p-1)/p} |c|^{(1-p)/p} |Lu| |w| \, dA$$

$$\leqslant \int_A |u|^{p-1} |c|^{(p-1)/p} |w|^{(p-1)/p} |Lu| |c|^{(1-p)/p} |w|^{1/p} \, dA$$

$$\leqslant \left(\int_A [|u|^{p-1} |c|^{(p-1)/p}]^{p/(p-1)} |w| \, dA \right)^{(p-1)/p}$$

$$\times \left(\int_A [|Lu| |c|^{(1-p)/p}]^p |w| \, dA \right)^{1/p}$$

$$= \left(\int_A |u|^p |c| |w| \, dA \right)^{(p-1)/p} \left(\int_A |Lu/c|^p |c| |w| \, dA \right)^{1/p}$$

$$= \left(\int_A |u|^p cw \, dA \right)^{(p-1)/p} \left(\int_A |L(u)/c|^p cw \, dA \right)^{1/p}.$$

209

Solving for

$$\left(\int_A |u|^p\, cw\, dA \right)^{1/p},$$

we obtain the conclusion of the lemma.

Our first maximum principle follows by a trivial application of the Cauchy–Riesz lemma by letting $p \to \infty$, noting that $cw > 0$.

Theorem 1 $(c: A \cup \Sigma \to R_-^1) \wedge (u \in C_L) \wedge (u = 0 \text{ a.e. on } \Sigma^2 \cup \Sigma^3) \Rightarrow$

$$\max_{A \cup \Sigma} |u| \leqslant \operatorname*{lub}_A |Lu/c|. \tag{12.4.4}$$

This can be regarded as an L^∞-norm a priori inequality for $u = 0$ a.e. on $\Sigma^2 \cup \Sigma^3$. We assume, of course, that L is of elliptic–parabolic type, and we assume the previously stated continuity conditions on coefficients of L as well as the definitions of Σ^2 and Σ^3. The beauty of the technique used in obtaining (12.4.4) is perhaps to be appreciated only by those with some experience. In any case, if we were to expound on the elegance of this technique, we would find it necessary to demonstrate the importance of (12.4.4), and we prefer to wait with this at least until a second maximum principle is proved. For now one should simply notice how and why the technique was applied as it was; i.e., one should notice the role of the condition that L be of elliptic–hyperbolic type coupled with the definition of Σ^1 and Σ^2, the need to let $v = u^2 + \delta$ with $\delta > 0$ and then examine the results as $\delta \to 0$, and finally the use of classic inequalities to get L^p-norm estimates and getting the maximum norm estimates through the Cauchy–Riesz lemma, letting $p \to \infty$.

5 A SECOND MAXIMUM PRINCIPLE

The second maximum principle is more like the familiar one for the Laplace operator since it concerns estimates for the homogeneous equation.

The set $R_-^1 \cup \{0\}$ is the set of nonpositive real numbers.

Lemma 1 $(p \geqslant 1) \wedge (w: A \cup \Sigma \to R_-^1 \cup \{0\}) \wedge (w \in C^2(A \cup \Sigma)) \wedge (L^*w + (p-1)cw > 0 \text{ in } A \cup \Sigma) \wedge (w = 0 \text{ a.e. in } \Sigma^3) \wedge (u \in C_L) \wedge (Lu = 0 \text{ in } A) \Rightarrow$

$$\left(\int_A |u|^p\, dA \right)^{1/p} \leqslant \left\{ \frac{\operatorname{lub}_{\Sigma^2} |bw|}{\min_{A \cup \Sigma} [L^*w + (p-1)cw]} \right\}^{1/p} \left(\int_{\Sigma^2} |u|^p\, ds \right)^{1/p}$$

$$+ \left\{ \frac{\operatorname{lub}_{\Sigma^3} |a^{ij}w_{x_i}n_j|}{\min_{A \cup \Sigma} [L^*w + (p-1)cw]} \right\}^{1/p} \left(\int_{\Sigma^3} |u|^p \right)^{1/p}. \tag{12.5.1}$$

Proof. From (12.3.7),

$$\int_A \{(u^2+\delta)^{p/2} L^*w - cw(u^2+\delta)^{(p/2)-1}[(1-p)u^2+\delta]\}\, dA$$

$$\leq \int_{\Sigma^2} bw(u^2+\delta)^{p/2}\, ds - \int_{\Sigma^3} (u^2+\delta)^{p/2} a^{ij} w_{x_i} n_j\, ds.$$

Letting $\delta \to 0$,

$$\int_A (u^2)^{p/2} [L^*w + (p-1)cw]\, dA \leq \int_{\Sigma^2} bw(u^2)^{p/2}\, ds - \int_{\Sigma^3} (u^2)^{p/2} a^{ij} w_{x_i} n_j\, ds.$$

$$(12.5.1a)$$

Again, for $p = 1$ this is the announced result. For $p > 1$, we simply use that for $A, B, C \geq 0$,

$$A \leq B + C \Rightarrow A^{1/p} \leq B^{1/p} + C^{1/p}.$$

We are now prepared to prove the second maximum principle.

Theorem 2 $(c: A \cup \Sigma \to R_-{}^1) \wedge (\exists w: A \cup \Sigma \to R_-{}^1 \ni L^*w + (p-1)w > 0)$
$\Rightarrow \forall u \in C_L \ni Lu = 0$,

$$\max_{A \cup \Sigma} |u| = \max_{\bar{\Sigma}^2 \cup \bar{\Sigma}^3} |u|. \qquad (12.5.2)$$

Proof If p is an even integer, we can use (12.3.7) with $\delta = 0$. Then

$$\min_{A \cup \Sigma} [L^*w + (p-1)cw] \int_A |u|^p\, dA \leq \operatorname*{lub}_{\Sigma^2} |bw| \int_{\Sigma^2} |u|^p\, ds$$

$$+ p \max_{\Sigma^3} |w| \int_{\Sigma^3} |u|^{p-1} |a^{ij} u_{x_i} n_j|\, ds$$

$$+ \operatorname*{lub}_{\Sigma^3} |a^{ij} w_{x_i} n_j - bw| \int_{\Sigma^3} |u|^p\, ds.$$

$$(12.5.3)$$

Case 1 $u = 0$ a.e. on $\bar{\Sigma}^3$.

There exist real numbers P_1, P_2 such that for large enough values of p

$$P_1 \leq \min_{A \cup \Sigma} [L^*w + (p-1)cw] \leq P_2 p$$

and, therefore,

$$\lim_{p \to \infty} \min_{A \cup \Sigma} [L^*w + (p-1)cw]^{1/p} = 1.$$

211

Then (12.5.2) follows from (12.5.3) by taking the pth root of both sides and letting $p \to \infty$ through even integers; notice that both integrals over Σ^3 in (12.5.3) are zero and that

$$\lim_{p \to \infty} (\operatorname*{lub}_{\Sigma^2} |bw|)^{1/p} = 1$$

by virtue of a preliminary proposition presented above.

Case 2 $u \neq 0$ a.e. on $\overline{\Sigma}^3$.

From the Schwarz–Hölder inequality,

$$\int_{\overline{\Sigma}^3} |u|^{p-1} |a^{ij} u_{x_i} n_j| \, ds \leqslant \left(\int_{\overline{\Sigma}^3} (|u|^{p-1})^{p/(p-1)} \, ds \right)^{(p-1)/p}$$

$$\times \left(\int_{\overline{\Sigma}^3} |a^{ij} u_{x_i} n_j|^p \, ds \right)^{1/p}$$

$$= \left(\int_{\overline{\Sigma}^3} |u|^p \, ds \right)^{(p-1)/p} \left(\int_{\overline{\Sigma}^3} |a^{ij} u_{x_i} n_j|^p \, ds \right)^{1/p}.$$

Then, from (12.5.3),

$$\min_{A \cup \Sigma} [L^* w + (p-1) cw] \int_A |u|^p \, dA \leqslant K_p \int_{\Sigma^2 \cup \Sigma^3} |u|^p \, ds, \qquad (12.5.4)$$

where

$$K_p = \operatorname*{lub}_{\Sigma^2} |bw| + p \max_{\Sigma^3} |w| \left(\int_{\overline{\Sigma}^3} |u|^p \, ds \right)^{1/p}$$

$$\times \left(\int_{\overline{\Sigma}^3} |a^{ij} u_{x_i} n_j|^p \, ds \right)^{1/p} + \operatorname*{lub}_{\Sigma^3} |a^{ij} w_{x_i} n_j - bw|.$$

Since $u \in C^1 (A \cup \Sigma)$, there is an $M \geqslant 0$ such that for every even integer p,

$$\left(\int_{\overline{\Sigma}^3} |u|^p \, ds \right)^{-1/p} \left(\int_{\overline{\Sigma}^3} |a^{ij} u_{x_i} n_j|^p \, ds \right)^{1/p} \leqslant M.$$

Then

$$K_p^{1/p} \leqslant [\operatorname*{lub}_{\Sigma^2} |bw| + p \max_{\overline{\Sigma}^3} |w| \, M + \operatorname*{lub}_{\Sigma^3} |a^{ij} w_{x_i} n_j - bw|]^{1/p}$$

$$\leqslant p^{1/p} [\operatorname*{lub}_{\Sigma^2} |bw| + M \max_{\overline{\Sigma}^3} |w| + \operatorname*{lub}_{\Sigma^3} |a^{ij} w_{x_i} n_j - bw|]^{1/p},$$

and

$$\limsup K_p^{1/p} \leqslant 1.$$

Let p_0 be such that

$$\min[L^*w + (p-1)cw] \geqslant 0$$

for every $p > p_0$. Then for every $p > p_0$, from (12.5.4)

$$\left(\min_{A \cup \Sigma}[L^*w + (p-1)cw]\right)^{1/p}\left(\int_A |u|^p ds\right)^{1/p} \leqslant K_p^{1/p}\left(\int_{\Sigma^2 \cup \Sigma^3} |u|^p ds\right)^{1/p},$$

and, for $p \to \infty$, we obtain the conclusion of the theorem. We have used that $\lim_{p\to\infty} K_p^{1/p}$ exists because all other factors have such limits.

A subtle but standard argument can be invoked to see that the maximum in (12.5.2) actually occurs on the boundary, not at an interior point. Thus (12.5.3) can be strengthened to say,

Corollary $(c: A \cup \Sigma \to R_-{}^1) \wedge (u \in C_L) \wedge (Lu = 0 \text{ in } A) \wedge (\exists x \in A \ni u(x) \neq 0) \Rightarrow$

$$\forall x \in A, \qquad |u(x)| < \max_{\bar{\Sigma}^2 \cup \bar{\Sigma}^3} |u(x)|. \tag{12.5.5}$$

Proof The function $w = 1$ satisfies Theorem 2. Of course, either

$$\max_{A \cup \Sigma} |u| = \max_{A \cup \Sigma} u \qquad \text{or} \qquad \max_{A \cup \Sigma} |u| = -\min_{A \cup \Sigma} u,$$

so u has in $A \cup \Sigma$ either a positive maximum or a negative minimum.

In A we have

$$Lu = a^{ij}u_{x_i x_j} + b^i u_{x_i} + cu = 0, \tag{12.5.6}$$

but, at a positive interior maximum point, $x \in A$,

$$u(x) > 0 \qquad \text{and} \qquad u_{x_i}(x) = 0 \tag{12.5.7}$$

for every $i = 1, \dots, n$. Then from (12.5.6)

$$a^{ij}(x)u_{x_i x_j}(x) > 0. \tag{12.5.8}$$

But then, if λ^i are eigenvalues of $A = (a^{ij})$ and v^i are scalar variables along the corresponding eigenvectors, we have

$$\lambda^i u_{v_i v_i} > 0. \tag{12.5.9}$$

But since $A = (a^{ij})$ is assumed to be positive semidefinite, $\lambda^i \geqslant 0$ for all $i = 1, \dots, n$.

Now if all $\lambda^i = 0$, then (12.5.9) cannot be so, and thus (12.5.8) cannot be so, and finally (12.5.7) cannot be valid, so there is no interior maximum. Otherwise, there exists $i = 1, \ldots, n$ such that $\lambda^i \neq 0$, and thus from (12.5.9)

$$u_{v_i v} \neq 0$$

so again x cannot be a point of maximum. The same argument of course, can be applied to $-u$ to reach the corollary.

This completes the derivation of a priori estimates necessary to prove the existence of L^p-weak solutions, but completion of the task will require a brief development of functional analysis up to a proof of the Hahn–Banach theorem on the continuation of continuous linear functionals. Once this development has been completed, we will expound the abstract existence principle and return to use the a priori inequalities to prove existence of L^p-weak solutions. Before turning our attention to these matters, we will prove the uniqueness of regular solutions and demonstrate that the a priori inequalities allow the computation of bounds of approximation errors for boundary-value problems related to elliptic–parabolic type.

The student should note that in (12.5.5) the maximum is found to occur on that portion of the boundary where data is to be specified. Such a result might well be all that one would want to know for a particular application.

13

Uniqueness of Regular Solutions and Error Bounds in Numerical Approximation

1 A COMBINED MAXIMUM PRINCIPLE

We now combine Theorem 1 and the corollary to Theorem 2 [Eqs. (12.4.4) and (12.5.5)] to give an a priori estimate that does not depend on either the condition $u = 0$ a.e. on $\Sigma^2 \cup \Sigma^3$ or the condition $Lu = 0$ in A. To this end, for every $v \in C_L(A)$, let

$$v = v_1 + v_2 \qquad (13.1.1a)$$

where

$$Lv_1 = 0 \quad \text{in} \quad A \qquad Lv_2 = L(v-v_1) = Lv \quad \text{in} \quad A$$

$$v_1 = v \quad \text{in} \quad \Sigma^2 \cup \Sigma^3 \qquad v_2 = 0 \quad \text{in} \quad \Sigma^2 \cup \Sigma^3,$$

and, once again, let $c: A \cup \Sigma \to R_-{}^1$. Then for every $v \in C_L$,

$$\max_{A \cup \Sigma} |v| = \max_{A \cup \Sigma} |v_1 + v_2| \leqslant \max_{A \cup \Sigma} |v_1| + \max_{A \cup \Sigma} |v_2|$$

$$\leqslant \max_{\bar{\Sigma}^2 \cup \bar{\Sigma}^3} |v_1| + \operatorname*{lub}_{A} |Lv_2/c|$$

$$= \max_{\bar{\Sigma}^2 \cup \bar{\Sigma}^3} |v| + \operatorname*{lub}_{A} |Lv/c| \qquad (13.1.1b)$$

215

The combined maximum principle is the inequality of the far left and far right members of (13.1.1b). It should be carefully noted that the validity of (13.1.1b) does not depend on any particular boundary-value problem being solved or on any conditions being satisfied that are directly related to it. It is asked only that $v \in C_L$ (solution regularity condition), that $c < 0$, tacitly that $a^{ij} \in C^2(A \cup \Sigma)$, $b^i \in C^1(A \cup \Sigma)$, $c \in C(A \cup \Sigma)$, and that L be of elliptic–parabolic type. With this noted, we proceed to use (13.1.1b) to prove uniqueness and to derive a formula for computation of error bounds.

2 UNIQUENESS OF REGULAR SOLUTIONS

For the linear boundary-value problem,

$$Lu = g \quad \text{in} \quad A,$$

$$u = f \text{ a.e.} \quad \text{on} \quad \Sigma^2 \cup \Sigma^3, \tag{13.2.1}$$

the uniqueness problem for regular solutions is simply to prove that

$$(u \in C_L(A)) \wedge (Lu = 0 \text{ in } A) \wedge (u = 0 \text{ a.e. on } \Sigma^2 \cup \Sigma^3) \Rightarrow u = 0 \text{ in } A \cup \Sigma.$$

Under these conditions Eq. (13.1.1b) gives that

$$\max_{A \cup \Sigma} |u| = 0, \tag{13.2.2}$$

whenever L is elliptic–parabolic type and $c < 0$. The result (13.2.2) completes our proof of uniqueness for regular solutions.

3 ERROR BOUNDS IN MAXIMUM NORM

Let $\{v^k\}$ be a finite set of approximation functions, $k = 1, 2, ..., n$, such that $v^k \in C_L$. We seek an approximation to the regular solution of (13.2.1), which we now know is unique if it exists. For this purpose we determine n real numbers c_k so that, where

$$u^n = \sum_{k=1}^{n} c_k v_k,$$

the following quadratic function is minimized,

$$Q(c_1, ..., c_n) = \int_A (Lu^n - g)^2 \, dA + \int_{\Sigma^2 \cup \Sigma^3} (f - u_n)^2 \, ds. \tag{13.3.1}$$

It should be noted that (3.5.2) is a special case of (13.3.1) where it is assumed that $g = 0$ and each element of the approximating family satisfies the differential equation. In that case we have

$$Lu^n = L\left(\sum c_k v^k\right) = \sum c_k Lv^k = 0 = g,$$

and the first term of the right member of (13.3.1) can be deleted.

Since Q is quadratic and $Q \geqslant 0$, there is one stationary vector (c_k) for Q, and it is a minimum. Therefore, we simply place $\partial Q/\partial c_k = 0$ and obtain a linear system of equations

$$\sum_{k=1}^{n} a^{hk} c_k = b^h, \qquad h = 1,\ldots,n \tag{13.3.2}$$

where

$$a^{hk} = \int_A Lv^k Lv^h \, dA + \int_{\Sigma^2 \cup \Sigma^3} v^h v^k \, ds \tag{13.3.3a}$$

and

$$b^h = \int_{\Sigma^2 \cup \Sigma^3} fv^k \, ds + \int_A gLv^k \, dA \tag{13.3.3b}$$

This system (of "normal equations of least squares") can be solved by elimination, or, if accuracy demands it, by accelerated successive replacements, an iterative method that converges for the normal equations of least squares because the coefficient matrix is symmetric positive-definite (see Chapter 3, Section 5). It turns out that this requires that the family $\{v^k\}$ of approximating functions be linearly independent (otherwise, the coefficient matrix may be only semidefinite) and that there exists a unique regular solution of (13.2.1), but these are no more than normal precaution would demand in the course of undertaking a numerical approximation.

Once the c_k's, and therefore u^n, have been determined, we let $v = u - u^n$ in (13.1.1), so that for the regular solution u of the boundary-value problem (13.2.1), we have

$$\max_{A \cup \Sigma} |u - u^n| \leqslant \max_{\bar{\Sigma}^2 \cup \bar{\Sigma}^3} |f - u^n| + \operatorname*{lub}_A \left|\frac{g - Lu^n}{c}\right|. \tag{13.3.4}$$

Since f, g, c—and now u^n and Lu^n—are known, this gives a method for computing a bound for the error of approximation function for regular solutions to (13.1.1) in maximum norm.

4 ERROR BOUNDS IN L^p-NORM

The right-hand side of (13.3.4) may be impractical to compute. Moreover, since our method of approximation utilizes an L^2-norm minimization, we will expect to be better satisfied with a bound (for the error function) in the same norm. Here we develop error bounds in L^p-norm, $p > 1$, but the student should remind himself throughout that the $p = 2$ case is likely to be the most important one given. It will be shown why, in this case, the error bounds turn out to be so very very computable.

We again separate every regular function $v \in C_L(A)$, as in (13.1.1a), into two components v_1 and v_2 so that v_1 satisfies the homogeneous equation and v_2 satisfies the homogeneous boundary condition $f = 0$ on $\Sigma^2 \cup \Sigma^3$. Then from Lemma 1 of Chapter 12, Section 4 [see (12.4.1)],

$$\left(\int_A |v_2|^p \, dA \right)^{1/p} \leqslant H_2(w) \left(\int_A |Lv_2|^p \, dA \right)^{1/p} \tag{13.4.1}$$

where w is any function such that

$$w \in C^2(A \cup \Sigma), \qquad w \leqslant 0, \qquad L^*w + (p-1)\,cw > 0 \text{ in } A \cup \Sigma, \tag{13.4.2a}$$

and

$$H_2(w) = p \, \frac{\max_{A \cup \Sigma} |w|}{\min_{A \cup \Sigma} [L^*w + (p-1)\,cw]}. \tag{13.4.2b}$$

From (12.5.1),

$$\left(\int_A |v_1|^p \, dA \right)^{1/p} \leqslant H_1(w) \left(\int_{\Sigma^2 \cup \Sigma^3} |v_1|^p \, ds \right)^{1/p} \tag{13.4.3}$$

where w is any function such that

$$w \in C^2(A \cup \Sigma), \qquad w: A \cup \Sigma \to R_-^{\,1} \cup \{0\}, \qquad w = 0 \text{ a.e. on } \Sigma^3, \tag{13.4.4a}$$

$$L^*w + (p-1)\,cw > 0 \text{ in } A \cup \Sigma, \tag{13.4.4b}$$

and $H_1(w)$ is largest of the two numbers

$$\left\{ \frac{\text{lub}_{\Sigma^2} |bw|}{\min_{A \cup \Sigma} [L^*w + (p-1)\,cw]} \right\}^{1/p} \quad \text{and} \quad \left\{ \frac{\text{lub}_{\Sigma^3} |a^{ij} w_{x_i} n_j|}{\min_{A \cup \Sigma} [L^*w + (p-1)\,cw]} \right\}^{1/p}. \tag{13.4.4c}$$

Now since $v = v_1 + v_2$, the triangle inequality condition on the L^p-norm, gives

$$\left(\int_A |v|^p\, dA\right)^{1/p} \leqslant \left(\int_A |v_1|^p\, dA\right)^{1/p} + \left(\int_A |v_2|^p\, dA\right)^{1/p,}$$

so that, from (13.4.1) and (13.4.3),

$$\left(\int_A |v|\, dA\right)^{1/p} \leqslant H(w)\left[\left(\int_{\Sigma^2 \cup \Sigma^3} |v_1|^p\, ds\right)^{1/p} + \left(\int_A |Lv_2|^p\, dA\right)^{1/p}\right]$$

where

$$H(w) = \max\{H_1(w), H_2(w)\}.$$

Of course, since $v_2 = 0$ a.e. in $\Sigma^2 \cup \Sigma^3$ and $Lv_1 = 0$ in A, we have for any $v \in C_L$ that

$$\left(\int_A |v|^p\, dA\right)^{1/p} \leqslant H(w)\left[\left(\int_{\Sigma^2 \cup \Sigma^3} |v|^p\, ds\right)^{1/p} + \left(\int_A |Lv|^p\, dA\right)^{1/p}\right]. \quad (13.4.5)$$

This, then, is a formula like that of (13.1.1b) which is in maximum (or supremum) norm but which is simpler in that it uses a coefficient $H(w) = 1$. If u is the unique regular solution of (13.2.1) and $u^n \in C_L(A)$ is to be regarded as an approximation to u, then an L^p-error bound for this approximation is evidently given by

$$\left(\int_A |u - u^n|^p\, dA\right)^{1/p} \leqslant H(w)\left[\left(\int_{\Sigma^2 \cup \Sigma^3} |f - u^n|^p\, ds\right)^{1/p} + \left(\int_A |g - Lu^n|^p\, dA\right)^{1/p}\right]. \quad (13.4.6)$$

5 COMPUTABLE BOUNDS FOR THE L^2-NORM OF AN ERROR FUNCTION

It is possible in several ways to obtain an inequality,

$$\int_A (u - u^n)^2\, dA \leqslant H(w)\left\{\int_A (g - Lu^n)^2\, dA + \int_{\Sigma^2 \cup \Sigma^3} (f - u^n)^2\, ds\right\}, \quad (13.5.1)$$

where $H(w)$ is determined in a manner which is similar to that for (13.4.6) in that one must first seek an admissible function w, and then compute certain bounds associated with w, before obtaining a valid level for $H(w)$. The equation (13.5.1) is not quite that of (13.4.6). The latter perhaps looks better when one thinks separately in terms of norms of errors in boundary-values and in deficiencies of the approximation u^n in satisfying the differential equation and boundary conditions, but it turns out that the right member of (13.4.6) is not as computable as that of (13.5.1).

The fact that this should be the case becomes apparent on noting that, except for the factor $H(w)$, the right member of (13.5.1) is the quantity, $Q(c_1, ..., c_n)$, to be minimized in the least-squares approximation procedure [see (13.3.1)], and least-squares theory provides that this quantity can always be computed very directly from knowledge of the values $c_1, ..., c_n$ which minimize Q—i.e., from the solution vector for the linear system (13.3.2), (13.3.3). Here it is given by

$$Q(c_1, ..., c_n) = \int_{\Sigma^2 \cup \Sigma^3} f^2 \, ds + \int_A g^2 \, dA - \sum_{h=1}^{n} c_h b_h, \qquad (13.5.2)$$

where the b_h's are components (13.3.3b) of the given vector on the right of the system of normal equations (13.3.2), and the c_h's are components of the solution vector. Of course, in practical computations the vector (c_h) will almost never be an exact solution of (13.3.2), but (13.5.2) can still be utilized if b_h is replaced by $b_h - r_h$ where (r_h) is the residual vector when the vector (c_h) is substituted in the set of normal equations.

EXERCISE 1 Prove the validity of (13.5.2).

An optimum choice of the functional $H(w)$ in (13.5.1) can often be made, and this also lends preference to the computation of L^2-norm error bounds.

14

Some Functional Analysis

1 GENERAL PRELIMINARIES

The Hahn–Banach theorem, the closed graph theorem, and the Banach–Steinhaus theorem are said to be the three pillars on which functional analysis rests; of these, our principle need will be for the Hahn–Banach theorem, and we will provide a proof for it. Functional analysis is essentially concerned with the study of the relations between functions when regarded as infinite-dimensional vectors and thus as the elements of (linear) vector spaces. Most of the activity in functional analysis centers around the consideration of a function space and its associated "adjoint space," the space of continuous linear functionals on it, and we find it necessary to develop these considerations as fully as our limitations allow. Since the norms of importance on these spaces are the L^p-norms, it is evident that integration theory plays some role. The student who feels uneasy about continuing now without some background in Lebesque integration will find many sources that have been specifically prepared to provide a quick presentation that is correct and easily readable. Among these are the following pamphlet size volumes:

W. W. Rogosinki, "*Volume and Integral*," Wiley (Interscience), New York, 1952

and

J. C. Burkill, "*The Lebesque Integral*," Cambridge Univ. Press, London and New York, 1958.

The undergraduate text,

J. Cronin-Scanlon, "*The Advanced Calculus*," Heath, Boston, Massachusetts, 1967

contains a thorough chapter on the subject. But for the level of completeness and generality the student will eventually feel a need to master, we must recommend an advanced text on real variables like that of Royden [39]. We believe, however, that at least the material of Chapter 14, if not even that of Chapter 15, can be read without feeling a really compelling need to accomplish this material first.

When we write here $f: A \to B$, then the domain of f *contains* A, and the range of f restricted to A is contained in B; in symbols $f \, r \, A \subset B$. These concepts will play a role in the Hahn–Banach theorem.

First, we introduce the concept of a vector space (linear space). As earlier, we let R^1 be the set of real numbers and let C^1 be the set of complex numbers. We use the symbol $\mathscr{F}$ as a generic symbol to stand for either.

Definition A vector space (linear space) over $\mathscr{F}$ is a system $\langle V, \oplus, \ominus, \odot, \bar{0}; \mathscr{F} \rangle$ where V is an abstract set, and the following seven axioms are satisfied:

1. $\bar{0} \in V$
2. $\oplus: V \times V \to V$
3. $\ominus: V \to V \wedge \ominus$ is (1–1)
4. $\forall x, y, z \in V, \; x \oplus (y \oplus z) = (x \oplus y) \oplus z \wedge x \oplus y = y \oplus x$
5. $\forall x \in V, \; x \oplus \bar{0} = x \wedge x \oplus (\ominus x) = \bar{0}$
6. $\odot: \mathscr{F} \times V \to V$
7. $\forall x, y \in V \wedge \forall \alpha, \beta \in \mathscr{F}$
       $1'.\quad \alpha \odot (x \oplus y) = (\alpha \odot x) \oplus (\alpha \odot y)$
       $2'.\quad (\alpha \odot \beta) \odot x = \alpha \odot (\beta \odot x)$
       $3'.\quad 1 \odot x = x.$

Semantics The symbols $\oplus$, $\ominus$, $\odot$ are called vector addition and subtraction, and multiplication of a vector by a scalar. The products of vector elements are not taken.

Agreements Having once made the distinction between $+$, $-$, $\cdot$, 0 and $\oplus$, $\ominus$, $\odot$, $\bar{0}$ we will drop the use of the last set of symbols, depending on context for this task. We will denote elements of V by lowercase Latin letters, $x, y, z, \ldots$, and elements of $\mathscr{F}$ by lowercase Greek letters, $\alpha, \beta, \delta, \ldots$. We will write

$$x - y \qquad \text{for} \qquad x + (-y)$$

and

$$\alpha x \quad \text{for} \quad \alpha \cdot x.$$

First Principles

1. $\forall x, y \in V, \ (x + y = x \Rightarrow y = 0) \wedge (0 \cdot x = 0)$
2. $\forall x, y \in V \wedge \forall \alpha \in \mathcal{F}, \ (\alpha x = \alpha y \wedge \alpha \neq 0) \Rightarrow (x = y)$
3. $\forall x \in V \wedge \forall \alpha, \beta \in \mathcal{F}, \ (\alpha x = \beta x \wedge x \neq 0) \Rightarrow (\alpha = \beta)$

EXERCISE 1 Prove the above first principles from the seven axioms.

Definition A linear manifold (subspace) in a linear space V over $\mathcal{F}$ is a set $M \subset V$ such that

(1) $M \neq \varnothing$ (the null set),
(2) $+ : M \times M \to M$, and
(3) $\cdot : \mathcal{F} \times M \to M$.

Proposition M is a linear manifold in V implies $\langle M, +, -, \cdot, 0; \mathcal{F} \rangle$ is a vector space.

Definition Let $S \subset V$. Then

$$\bar{S} = \left\{ x \in V \,\middle|\, \exists x_1, \ldots, x_n \in S \wedge \exists \alpha_1, \ldots, \alpha_n \in \mathcal{F} \ni x = \sum_{i=1}^{n} \alpha_i x_i \right\}$$

is said to be the space spanned by S.

Proposition $\bar{S}$ is a linear manifold in V.

EXERCISE 2 Prove the above propositions.

We have already used the concept of a function, which is usually taken to be a unique-valued relation with domain uniquely defined, but we include a definition of relation now for completeness.

Definition A relation R of a set P is any set of ordered couples of P, $(R \subset P \times P)$.

Agreement For any relation R of a set P, $(a, b) \in R$ will be denoted by $a \, R \, b$.

223

Definition The relation $<$ is said to partially order P if

1. $<$ is a relation;
2. $(P \times P \cap <) \neq \varnothing$, the null set;
3. $\forall a, b \in P, (a < a) \wedge ((a < b) \wedge (b < a)) \Rightarrow a = b$;
4. $\forall a, b, c \in P, ((a < b) \wedge (b < c)) \Rightarrow a < c$.

Definition A relation $<$ which partially orders P is said to completely[†] order P if for every $a, b \in P$, either $a < b$ or $b < a$.

Definition Let P be a set with partial ordering $<$, and let $A \subset P$. Then

1. $u \in P$ is an upper bound of A if for every $a \in A$, $a < u$;
2. m is a maximal element of A if for every $a \in A$, $m < a$ implies $m = a$.

Hausdorf Maximum Principle (HMP) Let $P \neq \varnothing$ be partially ordered by $<$. If every subset A of P which is completely ordered by $<$ has an upper bound in P, then P has a maximal element.

The prototype example is that of an open interval P and a closed interval P of real numbers (see Fig. 59). That not every partially ordered subset of an open interval P has a maximal element in P is noted by looking at P itself or any other interval A filling the upper end of P. However, for a closed interval P, the upper endpoint is an upper bound of any subset A of P and the upper endpoint is in P. Of course, it is also a maximal element of P.

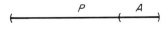

FIG. 59   An open interval P does not have the properties spoken of in the HMP; a closed interval P does.

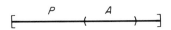

Comments In books on logic, the HMP is shown to be equivalent to the axiom of choice, which is usually one of the axioms of set theory—say in the Frankel system of set axioms. Here we will accept the HMP as an axiom without reference to the axiom of choice. Its role is that it is much more easily used to prove the basic theorems in analysis than is the axiom of choice.

We will next undertake to prove the Hahn–Banach theorem on the existence

---

† Perhaps the grammar here is bad, but it is standard.

of an extension of continuous linear functionals without change in norm. For this purpose we use only the HMP (to which "all men aver") and not a condition of separability. Using the separability property would enormously simplify the proof, and if the reader is familiar with a facile proof of the Hahn–Banach theorem it is with separability assumed, but it would not do for our application. The $\mathcal{L}^p$ spaces we wish to examine in order to prove the existence of $\mathcal{L}^p$-weak solutions for elliptic–parabolic type are certainly separable, but to prove this is a formidable technical task in analysis that we are not prepared to undertake.

2 THE HAHN–BANACH THEOREM, SUBLINEAR CASE

Definition A sublinear functional on the vector space $\langle V, +, \cdot, 0; R^1 \rangle$ over the real numbers is a function p such that

1. $p: V \to R^1$;
2. $\forall x, y \in V, \ p(x+y) \leqslant p(x) + p(y)$ (subadditive);
3. $\forall x \in V \land \forall \alpha \in R_+^1 \cup \{0\}, \ p(\alpha x) = \alpha p(x)$ (homogeneous).

Definition A linear functional on a vector space $\langle V, +, \cdot, -, 0; R^1 \rangle$ over the real numbers is a function L such that

1. $L: V \to R^1$;
2. $\forall x, y \in V, \ L(x+y) = L(x) + L(y)$ (additive);
3. $\forall x \in V \land \forall \alpha \in R_+^1 \cup \{0\}, \ L(\alpha x) = \alpha L(x)$ (homogeneous).

Theorem Let $\langle V, +, \cdot, -; R^1 \rangle$ be a real linear space. Let $M \subsetneqq V$ be a linear manifold. Let p be a sublinear functional on V. Let f be a linear functional on M such that for every $x \in M$ $f(x) \leqslant p(x)$. Then there exists a linear functional F on V such that (1) $Fr\,M = f$, (2) $\forall x \in V, \ F(x) \leqslant p(x)$.

Proof The proof presented here is very much the same as that given in [39]. It is extensive, taking up all the rest of the section.

(a)   Let $x_0 \in V - M$. Let $M_0 = M \cup \{x_0\}$; i.e., let M_0 be the linear manifold (subspace) spanned by the elements of $\overline{M \cup \{x_0\}}$ (see Section 1).

We now show that every $x \in M_0$ has one and only one expression of the form

$$x = y + \alpha x_0 \tag{14.2.1}$$

for some $y \in M$ and $\alpha \in R^1$. Clearly each $x \in M_0$ has one expression of this form. To show that it is unique we assume two such expressions to be available; i.e., we assume that $x \in M_0$ and

$$x = x_1 + \alpha_1 x_0 = x_2 + \alpha_2 x_0$$

for $x_1, x_2 \in M$ and $\alpha_1, \alpha_2 \in E^1$. Subtracting,

$$(x_1 - x_2) = (\alpha_2 - \alpha_1) x_0. \tag{14.2.2}$$

But $(x_1 - x_2) \in M$, and thus $(\alpha_2 - \alpha_1) x_0 \in M$. If $\alpha_1 \neq \alpha_2$, then

$$(\alpha_2 - \alpha_1)^{-1} (\alpha_2 - \alpha_1) x_0 = x_0 \in M,$$

which is a contradiction, so $\alpha_1 = \alpha_2$. But then by (14.2.2), $x_1 = x_2$.

(b) To the end of constructing a subdominant extension function F as required by the theorem, we define F_ε to be that function on M_0 such that

$$\forall x \in M \wedge \forall \alpha \in R^1, \ F_\varepsilon(x + \alpha x_0) = f(x) + \varepsilon \alpha. \tag{14.2.3}$$

We first show that F_ε is linear. Let $a, b \in M_0$ be given by

$$a = x + \alpha y_0, \qquad b = y + \beta x_0.$$

1. Additivity

$$\begin{aligned}
F_\varepsilon(a+b) &= F_\varepsilon[(x+y) + (\alpha+\beta) x_0] = f(x+y) + \varepsilon(\alpha+\beta) \\
&= f(x) + f(y) + \varepsilon(\alpha+\beta) = (f(x) + \varepsilon\alpha) + (f(y) + \varepsilon\beta) \\
&= F_\varepsilon(x + \alpha x_0) + F_\varepsilon(y + \beta x_0) \\
&= F_\varepsilon(a) + F_\varepsilon(b)
\end{aligned}$$

2. Homogeneity
   Let $\gamma \in R^1$.

$$\begin{aligned}
F_\varepsilon(\gamma a) &= F_\varepsilon(\gamma x + \gamma \alpha x_0) = f(\gamma x) + \gamma \alpha \varepsilon = \gamma f(x) + \gamma(\alpha \varepsilon) \\
&= \gamma(f(x) + \alpha \varepsilon) = \gamma F_\varepsilon(x + \alpha x_0) = \gamma F_\varepsilon(a)
\end{aligned}$$

(c) It is now to be shown that where a linear function F is defined by (13.2.3), we have that

$$\exists \varepsilon_0 \in R^1 \ni \forall x \in M, \ F_{\varepsilon_0}(x + \alpha x_0) \leqslant p(x + \alpha x_0). \tag{14.2.4}$$

If (14.2.4) is true, then with $\alpha = 1$,

$$f(x) + \varepsilon_0 = F_{\varepsilon_0}(x + 1 \cdot x_0) \leqslant p(x + x_0)$$

226

or

$$\varepsilon_0 \leqslant p(x+x_0) - f(x).$$

Also, putting $\alpha = -1$ and replacing x by $-x$,

$$f(-x) + \varepsilon_0 = F_{\varepsilon_0}(-x + (-1)x_0) \leqslant p(-x - x_0)$$

or

$$f(-x) - p(-x - x_0) \leqslant \varepsilon_0.$$

Thus (14.2.4) implies that

$$\forall x \in M, \qquad \exists \varepsilon_0 \in R^1 \ni f(-x) - p(-x - x_0) \leqslant \varepsilon_0 \leqslant p(x+x_0) - f(x).$$
$$(14.2.5)$$

Once it is shown that (14.2.4) and (14.2.5) are equivalent, we can demonstrate that (14.2.4) is true by showing that for any two values of x, the right side of the inequality in (14.2.5) is never smaller than the left, and then taking ε_0 to be any number between the infimum of the right member and the supremum of the left member. Toward this end we now show that (14.2.5) implies (14.2.4):

Case 1 $\alpha > 0$.

$$\forall x \in M, \qquad \alpha\varepsilon_0 \leqslant \alpha p(x+x_0) - \alpha f(x) = p(\alpha x + \alpha x_0) - f(\alpha x).$$

$$\forall y \in M \wedge \bar{x} = \alpha^{-1} y \in M,$$

$$\alpha\varepsilon_0 \leqslant p(\alpha\bar{x} + \alpha x_0) - f(\alpha\bar{x}) = p(y + \alpha x_0) - f(y).$$

Then

$$f(y) + \alpha\varepsilon_0 \leqslant p(y + \alpha x_0),$$

which is (14.2.4).

Case 2 $\alpha < 0$.

$$\forall x \in M, \qquad -\alpha f(-x) - (-\alpha)p(-x - x_0) \leqslant -\alpha\varepsilon_0$$

$$f(\alpha x) - p(\alpha x + \alpha x_0) \leqslant -\alpha\varepsilon_0.$$

$$\forall y \in M \wedge \bar{x} = \alpha^{-1} y \in M,$$

$$f(y) - p(y + \alpha x_0) = f(\alpha\bar{x}) - p(\alpha\bar{x} + \alpha x_0) \leqslant -\alpha\varepsilon_0.$$

Then

$$f(y) + \alpha\varepsilon_0 \leqslant p(y + \alpha x_0),$$

which is (14.2.4).

Case 3 $\alpha = 0$.

$$F_{\varepsilon_0}(y+\alpha x_0) = F_{\varepsilon_0}(y) = f(y) \leqslant p(y) = p(y+\alpha x_0).$$

This completes the demonstration that (14.2.4) is equivalent to (14.2.5). To demonstrate, then, that (14.2.4) is true we simply show that

$$\forall x_1, x_2 \in M, \qquad f(-x_1) - p(-x_1 - x_0) \leqslant p(x_2+x_0) - f(x_2).$$

But for $x_1, x_2 \in M$, we have

$$f(x_2) + f(-x_1) = f(x_2-x_1) \leqslant p(x_2-x_1)$$
$$= p(x_2+x_0-x_1-x_0)$$
$$\leqslant p(x_2+x_0) + p(-x_1 - x_0),$$

which is the desired inequality. From this

$$c = \sup_{x_1 \in M} \left(f(-x_1) - p(-x_1 - x_0)\right) \leqslant \inf_{x_2 \in M} \left(p(x_2+x_0) - f(x_2)\right) = G$$

and (14.2.5) [the equivalent of (14.2.4)] is established by selecting any ε_0 such that $c \leqslant \varepsilon_0 \leqslant G$.

(d)   We have shown that for every $\varepsilon \in R^1$, F_ε defined by (14.2.3) is a linear functional on M_0, and with ε_0 chosen satisfying (14.2.4), we have that

$$(F_{\varepsilon_0} r M = f) \wedge (\forall x \in M_0, F_{\varepsilon_0}(x) \leqslant p(x)).$$

This allows us to place an ordering on functionals with these properties by way of set inclusions of their domains. Once we have done this, the HMP will be applied to find a functional of this class which is maximal, and this will prove our theorem. Parts (b) and (c) of this proof have been devoted to showing simply that the set of such functionals is not empty, a necessary property for application of the HMP.

Definition Let P be the class of linear functionals g such that (see Fig. 60)

1. $\operatorname{dmn} g$ is a linear manifold in V,
2. $M \subset \operatorname{dmn} g$,
3. $g r M = f$,
4. $\forall x \in \operatorname{dmn} g, g(x) \leqslant p(x)$.

Definition A partial ordering is defined on P by the following: $\forall g, h \in P$, $g < h \Leftrightarrow (\operatorname{dmn} g \subset \operatorname{dmn} h) \wedge (h r \operatorname{dmn} g = g)$.

Suppose there is a nonempty set $A \subset P$ which is completely ordered by $<$. To apply the HMP, we must find $G \in P$ such that G is an upper bound of A.

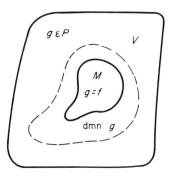

FIG. 60   An ordering is imposed on the class of functions P by a set inclusion of their domains.

Let

$$\mathscr{G} = \bigcup \{\operatorname{dmn} g \,|\, g \in A\},$$

and then let G be that function on $\mathscr{G}$ such that

$$\forall x \in \mathscr{G}, \; G(x) = \bigcup \{h(x) \,|\, (h \in A) \wedge (x \in \operatorname{dmn} h)\}.$$

It is evident that for every $h \in A$, $h < G$, but we must show that

$$G \in P. \tag{14.2.6}$$

The demonstration of this is divided into seven steps, the last three of which follow in a manner so well laid out here that their proofs are deleted:

1. $\mathscr{G} = \operatorname{dmn} G$ is a linear manifold in V.
   (i) $\forall x, y \in \mathscr{G}, \; x + y \in \mathscr{G}$.

   $(x \in \mathscr{G} \Rightarrow g \in A \ni x \in \operatorname{dmn} g) \wedge (y \in \mathscr{G} \Rightarrow \exists h \in A \ni y \in \operatorname{dmn} h)$.

Since A is completely ordered by $<$, we have $h < g$ or $g < h$. Without loss of generality, take $g < h$. Then $\operatorname{dmn} g \subset \operatorname{dmn} h$, and, therefore, $x \in \operatorname{dmn} h$. By definition of P, $\operatorname{dmn} h$ is a linear manifold in V, so $(x + y) \in \operatorname{dmn} h \subset \mathscr{G}$.

   (ii) $\forall x \in \mathscr{G} \wedge \alpha \in R^1, \; \alpha x \in \mathscr{G}$. $x \in \mathscr{G} \Rightarrow \exists g \in A \ni x \in \operatorname{dmn} g$. But the $\operatorname{dmn} g$ is a linear manifold in V, so $\alpha x \in \operatorname{dmn} g \subset \mathscr{G}$.
2. $M \subset \mathscr{G} = \operatorname{dmn} G$. Let $x \in M$. Since A is nonempty, there exists $g \in A \subset P$. But $M \subset \operatorname{dmn} g$, so $x \in \operatorname{dmn} g \subset \mathscr{G}$.
3. $\operatorname{Gr} M = f$. $\forall x \in M \wedge \forall h \in A, \; h(x) = f(x) \Rightarrow \{h(x) \,|\, h \in A\} = \{f(x)\}$. But $(\forall x \in M, \; G(x) = \bigcup \{h(x) \,|\, h \in A \wedge x \in \operatorname{dmn} h\} = \bigcup \{h(x) \,|\, h \in A\})$, so that $G(x) = \bigcup \{f(x)\} = \{f(x)\}$. We identify here $\{f(x)\}$ with f.
4. $G: \mathscr{G} \to R^1$. We have just shown that $\forall x \in M, \; G(x) \in R^1$, but we must show that $\forall x \in \mathscr{G}$

$$\bigcup \{h(x) \,|\, (h \in A) \wedge (x \in \operatorname{dmn} h)\} \in R^1.$$

229

Let $x \in \mathcal{G}$; then $\exists g \in A \ni x = \mathrm{dmn}\, g$. Let $h \in A$ such that $x \in \mathrm{dmn}\, h$; then either $h < g$ or $g < h$. Without loss of generality, let $h < g$. Then $g \, r \, \mathrm{dmn}\, h = h$ and $g(x) = h(x)$. Therefore,

$$G(x) = \bigcup \{h(x) \,|\, (h \in A) \wedge (x \in \mathrm{dmn}\, h)\}$$

$$= \bigcup \{g(x)\} = g(x) \in R^1.$$

The proof that $G \in P$ is concluded with the following:

5. $\forall x, y \in \mathcal{G},\ G(x+y) = G(x) + G(y)$,
6. $\forall x \in \mathcal{G} \wedge \forall \alpha \in R^1,\ G(\alpha x) = \alpha G(x)$,
and
7. $\forall x \in \mathcal{G},\ G(x) \leqslant p(x)$,
which can be shown in the same manner.

By the HMP, the set P has a maximal element F, and clearly F is the required extension of f; if not, we could, following the initial portion of this proof, extend F and obtain the element $F' \in P$ such that $F < F'$ and $F \neq F'$, which would contradict that F is maximal.

This completes our proof of this form of the Hahn–Banach theorem. Another less abstract and more useful form of it will be developed from the sublinear form just proved once the concept of a norm, and some associated facets of functional analysis, have been introduced.

3 NORMED SPACES AND CONTINUOUS LINEAR OPERATORS

Recall that $\mathcal{F}$ is a generic symbol for either R^1 or C^1.

Definition If $\langle V, +, -, \cdot, 0; \mathcal{F} \rangle$ is a linear space, then N is a norm on V if

1. $N : V \to R^1$;
2. $\forall x \in V,\ (N(x) \geqslant 0) \wedge ((N(x) = 0) \Leftrightarrow (x = 0))$;
3. $\forall x, y \in V,\ N(x+y) \leqslant N(x) + N(y)$; and
4. $\forall x \in V \wedge \forall \alpha \in \mathcal{F},\ N(\alpha x) = |a|\, N(x)$.

First Properites For N a norm on V, it is evident that

1. $\forall x, y \in V,\ N(x-y) = N(y-x)$;
2. $\forall x, y \in V,\ N(x-y) = 0 \Leftrightarrow (x = y)$;
3. $\forall x, y, z \in V,\ N(x-z) \leqslant N(x-y) + N(y-z)$.

230

Definition A linear operator L on a linear space V into a linear space V' is a function such that

1. $L: V \to V'$;
2. $\forall x, y \in V,\ L(x+y) = L(x) + L(y)$; and
3. $\forall x \in V \wedge \forall \alpha \in \mathscr{F},\ L(\alpha x) = \alpha L(x)$.

Definition Let V and V' be linear spaces over $\mathscr{F}$ with norms N and N', respectively. A function $f: V \to V'$ is said to be $(N - N')$ continuous at $x \in V$ if

$$\forall \varepsilon > 0,\ \exists \delta > 0 \ni \forall y \in V,\ N(x-y) < \delta \Rightarrow N'(f(x) - f(y)) < \varepsilon.$$

Definition A function $f: V \to V'$ is $(N - N')$ continuous on V if it is $(N - N')$ continuous for all $x \in V$.

We now prove four equivalent conditions for $(N - N')$ continuity. We elevate the results to the level of a theorem because of their importance, not because of the level of difficulty in proof.

Theorem If L is a linear operator on a linear space V with norm N into a linear space V' with norm N', the following are equivalent statements:

1. L in $(N - N')$ continuous on V,
2. $\exists M \in R^1 \ni \forall x \in V,\ N'(L(x)) \leqslant MN(x)$,
3. $\sup_{N(x) \leqslant 1} N'(L(x)) < \infty$
4. L is $(N - N')$ continuous at one point $x \in V$.

Proof (1)   We show that the first statement implies the second. Suppose

$$\forall M \in R^1,\ \exists x \in V \ni N'(L(x)) > MN(x),$$

and choose such an x. We note that $x \neq 0$ because $x = 0 \Rightarrow N'(L(x)) = MN(x) = 0$. Then

$$N'\left(\frac{L(x)}{N(x)}\right) = \frac{N'(L(x))}{N(x)} > M,$$

or

$$N'\left(L\left(\frac{x}{N(x)}\right)\right) > M.$$

Since the element $x/N(x)$ has norm 1,

$$\forall M \in R^1,\ \exists y \in V \ni N(y) = 1 \wedge N(L(y)) > M.$$

231

From this we will show that L is not continuous at $z = 0$. Let $0 \leqslant \varepsilon \leqslant 1$; we show that

$$\forall \delta > 0, \ \exists z \in V \ni N(z) < \delta \wedge N'(L(z)) > \varepsilon.$$

For $\delta > 0$, let $M \in R^1$ such that

$$1/M < \delta.$$

As stated, there exists $y \in V$ such that $N(y) = 1$ and $N'(L(y)) > M$; select then such a y and let $z = y/M$, we have

$$N'(z) = N'(L(y/M)) = N'\left(\frac{L(y)}{M}\right) = \frac{N'(L(y))}{M} > 1 \geqslant \varepsilon.$$

Of course,

$$N(z) = N(y/M) = (1/M) N(y) = (1/M) < \delta.$$

(2) We show that the second statement implies the third. From the second statement, for every $x \in V$ such that $N(x) \leqslant 1$,

$$N'(L(x)) \leqslant M N(x) \leqslant M.$$

(3) We show that the third statement implies the fourth. To this end, we will show that L is $(N - N')$ continuous at the particular point $x = 0$. Thus, for every $\varepsilon > 0$ we must produce a $\delta > 0$ such that for every $y \in V$,

$$N(y) < \delta \Rightarrow N'(L(y)) < \varepsilon.$$

For this purpose, let

$$M = \sup_{N(x) \leqslant 1} N'(L(x))$$

and let $y' \in V$ such that $N(y') = 1$. Choose $0 < \delta < \varepsilon/M$. Then

$$\delta N'(L(y')) \leqslant \delta M < \varepsilon.$$

Letting $N(y) < \delta$ and $y' = y/\delta$,

$$N(y') = N(y/\delta) = (1/\delta) N(y) \leqslant \delta/\delta = 1$$

Then

$$N(y') = N'(L(\delta y')) = \delta N'(L(y')) < \varepsilon.$$

(4) We show that the fourth statement implies the first, and this completes a cycle of implications necessary to establish the theorem (see Fig. 61). Let L be continuous at $x_0 \in V$ and let $x \in V$ be such that $x \neq x_0$. We establish that

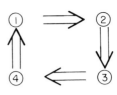

FIG. 61 The cycle of implications in the theorem on continuity of linear operators.

L is continuous at x; i.e., we show that

$$\forall \varepsilon > 0, \ \exists \delta > 0 \ni \forall z \in V, \ N(x-z) < \delta \Rightarrow N'(L(x) - L(z)) < \varepsilon.$$

For this purpose, let

$$N((x-z+x_0) - x_0) = N(x-z) < \delta;$$

then

$$N'(L(x+x_0-z) - L(x_0)) < \varepsilon.$$

But

$$L(x+x_0-z) - L(x_0) = L(x+x_0-z-x_0) = L(x-z),$$

so

$$N'(L(x-z)) < \varepsilon.$$

4 BANACH SPACES

Let Ω be the set of positive integers.

Definition If $\langle X, +, -, \cdot, 0; \mathscr{F}, N \rangle$ is a normed linear space, then

1. A sequence $s: \Omega \to X$ is N-Cauchy if $\forall \varepsilon > 0, \ \exists M \in \Omega \ni \forall m, n \in \Omega$

$$m, n > M \Rightarrow N(s_m - s_n) < \varepsilon.$$

2. A sequence $s: \Omega \to X$ converges N-wise to an element $x \in X$ if $\forall \varepsilon > 0,$ $\exists M \in \Omega \ni \forall n \in \Omega$

$$n > M \Rightarrow N(x - S_n) < \varepsilon.$$

We write $\{S_n\} \to x$.

3. $\langle X, +, -, 0; \mathscr{F}, N \rangle$ is said to be N-complete if for every N-Cauchy sequence $S: \Omega \to X$, there exists $x \in X$ such that $\{S_n\} \to x$.

Definition A Banach space is a normed linear vector space $\langle X, +, -, \cdot, 0;$ $\mathscr{F}, N \rangle$ which is N-complete.

Definition Let $\mathcal{X}_i = \langle X_i, +, -, \cdot, 0; \mathcal{F}, N_i \rangle$, $i = 1, 2$, be two normed linear spaces over $\mathcal{F}$, and denote by $[\mathcal{X}_1, \mathcal{X}_2]$ the set of all $N_1 - N_2$ continuous linear operators on $\mathcal{X}_1$ into $\mathcal{X}_2$.

Definition For every $A, B \in [\mathcal{X}_1, \mathcal{X}_2]$, let $A + B$ be that element $C \in [\mathcal{X}_1, \mathcal{X}_2]$ such that

$$\forall x \in X_1, \qquad C(x) = A(x) + B(x).$$

Definition For every $A \in [\mathcal{X}_1, \mathcal{X}_2]$ and for every $\alpha \in \mathcal{F}$, let αA be that element $C \in [\mathcal{X}_1, \mathcal{X}_2]$ such that

$$\forall x \in X_1, \qquad C(x) = \alpha A(x).$$

We write $(\alpha A)(x) = \alpha A(x)$.

Definition (Banach norm) For every $A \in [\mathcal{X}_1, \mathcal{X}_2]$, let

$$N_3(A) = \sup_{N_1(x) \leqslant 1} N_2(A(x)) = \sup\{N_2(A(x)) \mid (x \in X_1) \wedge (N_1(x) \leqslant 1)\}$$

Definition For $A \in [\mathcal{X}_1, \mathcal{X}_2]$ let $-A = (-1) \cdot A$. Let 0 be that element $C \in [\mathcal{X}_1, \mathcal{X}_2]$ such that

$$\forall x \in X_1, \qquad C(x) = 0 \in X_2.$$

Theorem The set of continuous linear operators $[\mathcal{X}_1, \mathcal{X}_2]$, with $+, -, \cdot, 0$ defined as above and the Banach norm adopted, is a normed linear space over $\mathcal{F}$,

$$\mathcal{X}_3 = \langle [\mathcal{X}_1, \mathcal{X}_2], +, -, \cdot, 0; \mathcal{F}, N_3 \rangle,$$

and, moreover, if $\mathcal{X}_2$ is N_2-complete, then $\mathcal{X}_3$ is N_3-complete.

The proof of completeness, which is nontrivial, is deleted here because it will not be needed for proof of the sufficiency of the Fichera abstract existence principle which is to be developed here in order to establish an existence theorem for elliptic–parabolic type.

The next definition concerns a special case of great importance where the space $\mathcal{X}_2$ in the above is simply the field of scalars $\mathcal{F}$ over which $\mathcal{X}_1$ is taken.

Definition Let $\mathcal{X} = \langle X, +, \cdot, -, 0; \mathcal{F}, N_1 \rangle$ be a normed linear space. Then the adjoint (or conjugate or dual) space is

$$\mathcal{X}^* = \langle [\mathcal{X}, \mathcal{F}], +, -, \cdot, 0; \mathcal{F}, N_3 \rangle$$

where $\forall A \in [\mathscr{X}, \mathscr{F}] \wedge \forall x \in X,$

$$N_3(A) = \sup_{N_1(x) \leqslant 1} N_2(A(x)) = \sup_{N_1(x) \leqslant 1} |A(x)|.$$

The adjoint space $\mathscr{X}^*$ is a normed linear space.

The following, then, is a corollary to the theorem above.

Corollary Let

$$\mathscr{X}^* = \langle [\mathscr{X}, \mathscr{F}], +, \cdot, 0; \mathscr{F}, N \rangle,$$

where

$$\mathscr{X} = \langle X, +, \cdot, 0; \mathscr{F}, N_1 \rangle$$

is a normed linear space over $\mathscr{F}$. Then $\mathscr{X}^*$ is a Banach space under the norm

$$N(x^*) = \sup_{N_1(x) \leqslant 1} |x^*(x)|$$

for $x \in X$ and $x^* \in [\mathscr{X}, \mathscr{F}]$.

5 THE HAHN-BANACH THEOREM FOR NORMED SPACES

We now return to the promised less abstract and more useful version of the Hahn–Banach theorem.

Theorem Let $\mathscr{X} = \langle X, +, \cdot, -, 0; R^1, N \rangle$ be a normed linear space over R^1. Let M be a proper linear manifold in $\mathscr{X}$. Let $m \in M^* = [M, R^1]$. Then there exists $x^* \in \mathscr{X}^* = [\mathscr{X}, E^1]$ such that $x^* r M = m \wedge N^*(x^*) = N^*(m)$ (see Fig. 62).

Proof For every $x \in \mathscr{X}$, let

$$p(x) = N^*(m) N(x) = \sup_{N(y) \leqslant 1} |m(y)| N(x).$$

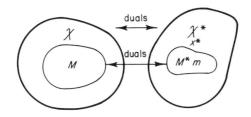

FIG. 62   Spaces, subspaces, and their dual spaces.

Then $p(x)$ is a norm and hence a sublinear functional. We must show that for every $x \in M$, $m(x) \leqslant N^*(m) N(x) = p(x)$. Now

$$\frac{1}{N(x)} m(x) = m\left(\frac{x}{N(x)}\right) \leqslant \sup_{N(y) \leqslant 1} |m(y)| = N^*(m),$$

which implies that $m(x) \leqslant N^*(m) N(x)$. The earlier form (see Section 2) of the Hahn–Banach theorem assures the existence of a linear functional x^* on $\mathcal{X}$ such that

$$x^*(x) \leqslant p(x) = N^*(m) N(x) \qquad \text{and} \qquad x^* r M = m.$$

Let $x \in \mathcal{X}$ and note that $N(-x) = N(x)$. Then

$$x^*(-x) = -x^*(x) \leqslant N^*(m) N(-x) = N^*(m) N(x).$$

Hence

$$\pm x^*(x) \leqslant N^*(m) N(x).$$

Therefore,

$$|x^*(x)| \leqslant N^*(m) N(x).$$

Now,

$$N^*(x^*) = \sup_{N(x) \leqslant 1} |x^*(x)| \leqslant N^*(m) < \infty.$$

Thus, x^* is continuous and $x^* \in \mathcal{X}^*$, and we already have that $N^*(x^*) \leqslant N^*(m)$
Since $x^* = m$ on the domain of m, $N^*(m) \leqslant N^*(x^*)$. Together, these yield simply that

$$N^*(x^*) = N^*(m).$$

This completes our proof.

As stated earlier we will use the Hahn–Banach theorem on $\mathcal{L}^p$ spaces, only, and these spaces are separable. Also, as stated earlier, the theorem can be stated and proved in a much simpler way if this property is assumed, but then one would have to take up a separate and difficult proof for the separability of $\mathcal{L}^p$-spaces. We give the statement of the theorem as it is given in Vol. 1 of Kolmogorov and Fomin [26]. The student may wish to consult the proof given there.

Theorem Every continuous linear functional f defined on a linear subspace G of a separable normed linear space E can be extended to the entire space with preservation of norm; i.e., it is possible to construct a linear functional $F(x)$ such that

1. $F(x) = f(x),\ x \in G$;
2. $N_E(F) = N_G(f)$.

We conclude this section with an example intended to demonstrate that not all linear functionals are continuous. Let $S = C[0, 1]$, the set of all functions that are continuous on the closed interval $[0, 1]$. Then, of course,

$$\langle S, +, \cdot, 0; R^1 \rangle$$

is a linear space. We can define

$$N(f) = \int_0^1 |f|,$$

and

$$L: S \to R^1 \qquad \text{by} \quad Lf = f(0).$$

Then $\langle S, +, -, 0; R^1, N \rangle$ is a normed linear space (separable, as it happens, but not complete) and L is linear. We show that L is not bounded (see property 3 of the theorem in Section 3). For every $n \in \Omega$, let

$$f_n(x) = \begin{cases} 2n - 2n^2 x & \text{for} \quad 0 \leqslant x \leqslant 1/n \\ 0 & \text{for} \quad 1/n \leqslant x \leqslant 1. \end{cases}$$

(See Fig. 63.) Then for every n,

$$N(f_n(x)) = 1 \wedge |Lf_n| = 2n.$$

Hence

$$\sup_{N(f) \leqslant 1} |Lf| = \infty,$$

and L is not continuous, though it is linear.

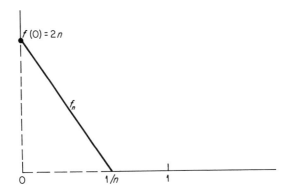

FIG. 63   The linear functional $Lf = f(0)$ is not bounded.

6 FACTOR SPACES

Let $\mathscr{X}$ be a normed linear space and let M be a linear manifold in $\mathscr{X}$. For every $x, y \in \mathscr{X}$, we write

$$x \equiv y \bmod M \qquad \text{if} \quad x - y \in M.$$

This is an equivalence relation on $\mathscr{X}$.

Definition The set of equivalence classes in $\mathscr{X}$ formed with reference to a subspace M by the relation $x \equiv y \bmod M$ is called the factor space of M in $\mathscr{X}$ and is written $\mathscr{X}/M$.

Proposition If addition and multiplication are defined in "the natural way" on $\mathscr{X}/m$, it is a linear space with zero element (additive identity) M. "The natural way" of defining addition and multiplication in $\mathscr{X}/M$ is as follows where $[x]$ denotes an equivalence class to which x belongs:

1. $[x] + [y] = [x+y]$,
2. $\alpha[x] = [\alpha x]$.

Justification

$(1) \quad [u] = [x], \ [v] = [y] \Rightarrow u \equiv x \bmod M, \ v \equiv y \bmod M$

$\qquad \qquad \qquad \qquad \Rightarrow u - x \in M, \ v - y \in M$

$\qquad \qquad \qquad \qquad \Rightarrow (u-x) + (v-y) \in M \ (M \text{ is a linear manifold})$

$\qquad \qquad \qquad \qquad \Rightarrow (u+v) - (x+y) \in M$

$\qquad \qquad \qquad \qquad \Rightarrow (u+v) \equiv (x+y) \bmod M \Rightarrow [u+v] = [x+y]$

$(2) \quad [u] = [x] \Rightarrow u - x \in M$

$\qquad \qquad \qquad \Rightarrow \alpha(u-x) \in M \ (M \text{ is a linear manifold})$

$\qquad \qquad \qquad \Rightarrow \alpha u - \alpha x \in M$

$\qquad \qquad \qquad \Rightarrow \alpha u = \alpha x \bmod M \Rightarrow [\alpha u] = [\alpha x].$

Definition Let $\mathscr{X}$ be a normed linear space with norm N. Let $F \subset \mathscr{X}$. Then

1. x is an N-limit point of F if

$$\forall \varepsilon > 0, \ \exists y \in F \ni \forall x \neq y, \ N(y-x) < \varepsilon;$$

and
2. F is said to be closed if it contains all its limit points.

238

Theorem Let $\mathscr{X}$ be a normed linear space and let M be a closed linear manifold in $\mathscr{X}$. For $[x] \in \mathscr{X}/M$, let

$$\overline{N}(x) = \inf_{y \in [x]} N(y).$$

Then $\overline{N}$ is a norm on $\mathscr{X}/M$, and if $\mathscr{X}$ is N-complete, then $\mathscr{X}/M$ is $\overline{N}$-complete.

Proof The conclusion concerning completeness is not needed here and will not be proved. We will prove only that $\overline{N}$ is a norm on $\mathscr{X}/M$.

Let $[x], [y] \in \mathscr{X}/M$. Then

$$\overline{N}([x] + [y]) = \inf_{z \in [x]+[y]} N(z) = \inf_{a \in [x], b \in [y]} N(a+b)$$

$$\leqslant \inf_{a \in [x], b \in [y]} (N(a) + N(b))$$

$$= \inf_{a \in [x]} N(a) + \inf_{b \in [y]} N(b) = \overline{N}(x) + \overline{N}(y).$$

Thus $\overline{N}$ is a subadditive functional. The rest of the norm properties follow trivially.

7 STATEMENT (ONLY) OF THE CLOSED GRAPH THEOREM

Definition Let $(\mathscr{X}_1, N_1)$, $(\mathscr{X}_2, N_2)$ be normed linear spaces. A mapping f with $\mathrm{dmn} f \subset \mathscr{X}_1$ and $\mathrm{rng} f \subset \mathscr{X}_2$ is closed if for every $S: \Omega \to \mathrm{dmn} f$,

$$(\{S_n\} \to x \in \mathscr{X}_1) \wedge (\{f(S_n)\} \to y \in \mathscr{X}_2) \Rightarrow (x \in \mathrm{dmn} f) \wedge (y = f(x)).$$

The closed graph theorem simply states that every closed linear map (graph) is continuous, to wit:

Theorem Let $(\mathscr{X}_1, N_1)$ and $(\mathscr{X}_2, N_2)$ be Banach spaces and let $f: \mathscr{X}_1 \to \mathscr{X}_2$.

If f is closed and linear, then f is $N_1 - N_2$ continuous. The proof of this fundamental theorem will not be undertaken here because it is used in proving only the necessity, not the sufficiency, of Fichera's abstract existence principle.

15

Existence of $\mathscr{L}^p$-Weak Solutions

1 A FIRST FORM OF THE ABSTRACT EXISTENCE PRINCIPLE

Here the functional value $N(x)$ of a norm of x will be denoted $\|x\|$. Let V be a linear manifold and let B_1, B_2 be Banach spaces. Let

$$M_1 : V \to B_1 \qquad \text{and} \qquad M_2 : V \to B_2$$

be linear operators; let Φ and Ψ denote vectors of the adjoint spaces $B_1{}^*$ and $B_2{}^*$ (spaces of continuous linear functionals on B_1 and B_2), respectively (see Fig. 64). For every $v \in V$, consider the functional equation

$$\Phi[M_1(v)] = \Psi[M_2(v)], \qquad (15.1.1)$$

where Φ is a given vector; we wish to establish existence of Ψ under some conditions that will be useful to us in the theory of boundary-value problems of elliptic–parabolic type. The relation of this theory to (15.1.1) may be quite enigmatic at this time; it has to do with the existence of $\mathscr{L}^p$-weak solutions which was introduced in somewhat general terms in Chapter 5, Section 8 [compare (15.1.1) with (5.8.6b)] and which will be taken up in these specific terms in Sections 2 and 3 of this chapter.

Theorem For every Φ, a solution Ψ of the functional equation (15.1.1) exists if and only if

$$\exists K \in R^1 \ni \forall v \in V, \qquad \|M_1(v)\| \leqslant K \|M_2(v)\|. \qquad (15.1.2)$$

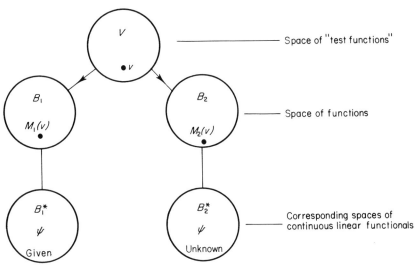

FIG. 64   Schematic, abstract existence principle.

Moreover, if (15.1.2) is satisfied, a solution Ψ exists satisfying the inequality

$$\|\Psi\| \leqslant K\|\Phi\|, \tag{15.1.3}$$

and every other solution is obtained by adding a functional Ψ' such that

$$\forall v \in V, \qquad \Psi'[M_2(v)] = 0. \tag{15.1.4}$$

Proof We will include here proofs of both the sufficiency and necessity of (15.1.2), but since we did not include a proof of the closed graph theorem in Chapter 14, the necessity portion of the proof will not be self-contained here. We make no apology for this procedure since in our application to existence of boundary-value problems, a sufficient condition for existence seems to be enough. The necessity assures us that no better condition can be found at this level of generality. Since there is essentially no "escape" from a necessary and sufficient condition, one can feel that if he masters this theorem and how to use it, he has gone a long way toward mastering the theory of partial differential equations; at least this is so insofar as weak solutions of linear boundary-value problems are concerned.

(a) *Condition (15.1.2) is sufficient* Let $M_2(V)$ be the image of V under M_2. For each $\omega_2 \in M_2(V)$, consider v such that $M_2(v) = \omega_2$ and the corresponding element of B_1, namely $M_1(v)$ (see Fig. 65). We show that $M_1(v)$ is uniquely

241

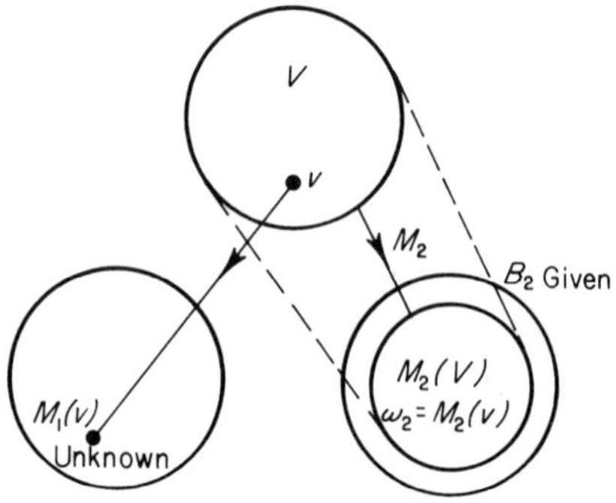

FIG. 65   The M_2 image of V.

determined by the choice of $\omega_2 \in M_2(V)$. For every v, $v' \in V$ such that $M_2(v) = M_2(v')$, (15.1.2) implies that

$$\|M_1(v-v')\| \leqslant \|M_2(v-v')\| = 0$$

or

$$M_1(v) = M_1(v').$$

Thus we can determine a functional ψ on $M_2(V)$ as follows

$$\psi(\omega_2) = \Phi[M_1(v)]. \tag{15.1.5}$$

Since M_1, M_2, and Φ are linear, it follows that ψ is a linear functional on $M_2(V)$:

$$\psi(\omega_2 + \omega_2{}^*) = \Phi[M_1(v+v^*)] = \Phi[M_1(v) + M_2(v^*)]$$
$$= \Phi[M_1(v)] + \Phi[M_2(v^*)] = \psi(\omega_2) + \psi(\omega_2{}^*).$$

Furthermore, from (15.1.2), ψ is continuous

$$\|\psi(\omega_2)\| = |\psi(\omega_2)| = |\Phi[M_1(v)]| \leqslant \|\Phi\| \|M_1(v)\|$$
$$\leqslant K\|\Phi\| \|M_2(v)\| = K\|\Phi\| \|\omega_2\| \tag{15.1.6}$$

because $K\|\Phi\| \in R^1$. From the Hahn–Banach theorem, this continuous linear functional can be extended to B_2.

Let Ψ be such an extension; it is a solution then of (15.1.1). Moreover, according to that theorem,

$$\|\Psi\| = \|\psi\| = \sup_{\|\omega_2\| \leqslant 1} |\psi(\omega_2)| \leqslant K\|\Phi\|. \tag{15.1.7}$$

This, of course, is the promised condition (15.1.3).

If Ψ_0 is orthogonal to $M_2(V)$, i.e., if

$$\Psi_0[M_2(v)] = 0,$$

it is evident that $\Psi + \Psi_0$ is another solution of (15.1.1). Moreover, if Ψ and Ψ' are solutions of (15.1.1), $\Psi - \Psi'$ is orthogonal to $M_2(v)$. This is the last statement (15.1.4) in the theorem.

(b) *Condition (15.1.2) is necessary* Instead of considering the whole of spaces B_1 and B_2, we now confine our attention to linear manifolds $M_1(V)$, $M_2(V)$ and to the continuous linear functionals defined on them. We indicate by $M_1{}^*$, $M_2{}^*$ the adjoint spaces of $M_1(V)$, $M_2(V)$, respectively. Then, for every $\Phi \in M_1{}^*$ we assume that there exists a unique solution $\Psi \in M_2{}^*$ of (15.1.1). This defines a transformation

$$\Psi = T(\Phi). \tag{15.1.8}$$

Suppose $\Psi_1 = T(\Phi_1)$ and $\Psi_2 = T(\Phi_2)$. It then follows that

$$(\Phi_1 + \Phi_2)[M_1(v)] = \Phi_1[M_1(v)] + \Phi_2[M_2(v)] = \Psi_1[M_2(v)] + \Psi_2[M_2(v)]$$

$$= (\Psi_1 + \Psi_2)[M_2(v)]$$

or

$$T(\Phi_1 + \Phi_2) = \Psi_1 + \Psi_2 = T(\Phi_1) + T(\Phi_2),$$

so T is linear.

We will show that T is closed. Let $\{\Phi_n\}$ be a sequence of elements in $M_1{}^*$ which converge to an element Φ and let $\{\Psi_n\} = \{T(\Phi_n)\}$ converge to Ψ. Since

$$\forall v \in V, \ \Phi_n[M_1(v)] = \Psi_n[M_2(v)],$$

letting $n \to \infty$, we have

$$\forall v \in V, \ \Phi[M_1(v)] = \Psi[M_2(v)].$$

Thus $\Psi = T(\Phi)$, and T is closed. It follows, of course, from the closed graph theorem that T is continuous. Therefore, we find that $T: M_1{}^* \to M_2{}^*$ defined by (15.1.8) is a continuous linear transformation. It can be considered onto without loss by replacing $M_2{}^*$ with the set of images $T(M_1{}^*)$.

Let M_1^{**} be the set of continuous linear functionals on $M_1{}^*$ and let M_2^{**} be the set of continuous linear functionals on $T(M_1{}^*)$, a restriction of $M_2{}^*$. Consider now a transformation $T': M_2^{**} \to M_1^{**}$ defined by

$$T'(\Psi') = \Phi'$$

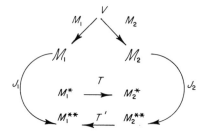

FIG. 66 Second conjugate spaces of M_1, M_2 images of V and canonical mappings J_1, J_2.

where $\Phi'(\Phi) = \Psi'(T(\Phi))$. Since T' is the conjugate of T, we have $\|T\| = \|T'\|$; i.e., we have

$$\|\Phi'\| = \|T'(\Psi')\| \leqslant K \|\Psi'\|.$$

Letting $J_1: \mathscr{M}_1 \to M_1^{**}$, $J_2: \mathscr{M}_2 \to M_2^{**}$ be the canonical maps (see Fig. 66), which pair the natural correspondents of $\mathscr{M}_1$ and $\mathscr{M}_2$ with M_1^{**} and M_2^{**}, we have $\|J\| = 1$. Then we can put

$$T'(J_2[\mathscr{M}_2(v)]) = \Phi' \qquad \text{or} \qquad J_1(\mathscr{M}_1(v)) = T'[J_2(\mathscr{M}_2(v))]$$

where for every $\Phi \in M_1^*$,

$$\Phi'(\Phi) = J_2(\mathscr{M}_2(v))[T(\Phi)] = T(\Phi)[\mathscr{M}_2(v)] = \Psi[\mathscr{M}_2(v)] = \Phi[\mathscr{M}_1(v)]$$
$$= J_1[\mathscr{M}_1(v)](\Phi).$$

Therefore,

$$T'(J_2[\mathscr{M}_2(v)]) = \Phi' = J_1(\mathscr{M}_1(v)),$$

and hence,

$$\|\mathscr{M}_1(v)\| = \|J_1[\mathscr{M}_1(v)]\| = \|T'(J_2[\mathscr{M}_2(v)])\| \leqslant K \|J_2[\mathscr{M}_2(v)]\|$$
$$= K \|M_2(v)\|.$$

Thus the proof is complete.

2 FUNCTION SPACES $\mathscr{L}^p$ AND $\mathscr{L}^{p/(p-1)}$; RIESZ REPRESENTATION

The class of functions on a region $A \subset E^n$ whose absolute value to the pth power, $p \geqslant 1$, is Lebesque integrable is denoted by $L^p(A)$. Two functions in $L^p(A)$ are taken to be equivalent if they differ only on a set of measure zero, and the set of equivalence classes in $L^p(A)$ under the natural definition of addition and multiplication by R^1 (see Chapter 14), form a linear space, in fact a Banach space which we denote by $\mathscr{L}^p(A)$. Thus, although we have avoided the distinction until now in order to avoid confusion, $L^p(A)$ is simply a set of functions while $\mathscr{L}^p(A)$ is a linear space. Not only that, but the set upon which the space $\mathscr{L}^p(A)$ is constructed is the set of equivalence classes on $L^p(A)$ which identify those functions which have the same Lebesgue integral over A. This should be kept in mind when interpreting any uniqueness statements for boundary-value problems that arise from application of the abstract existence principle to $\mathscr{L}^p$ function spaces.

We utilize without proof that the space of continuous linear functionals on $\mathscr{L}^p$ can be identified with $\mathscr{L}^q$, where

$$\frac{1}{p} + \frac{1}{q} = 1$$

[i.e., that $(\mathscr{L}^p)^*$ is isomorphic, isometric to $\mathscr{L}^q$], and, moreover, that

$$\forall \Phi \in (\mathscr{L}^p(A))^*, \ \exists f \in \mathscr{L}^p(A) \ni \Phi[v] = \int_A fv \, dA \qquad (15.2.1)$$

The statement (15.2.1) is the Riesz representation theorem. It allows us, once the existence of certain functionals on $\mathscr{L}^p(A)$ has been demonstrated, to assert the existence of functions belonging to $\mathscr{L}^p(A)$ that are associated with it by (15.2.1). But (15.2.1) has the appearance of an inner product, and it happens that continuous linear functionals play a role in $\mathscr{L}^p$ spaces which is a clear extension of the role of inner products in $\mathscr{L}^2$ spaces, so we adopt the notation $\langle \varphi, v \rangle$ for $\Phi[v]$. Thus the functional equation (15.1.1) is now to be written

$$\langle \varphi, M_1(v) \rangle = \langle \psi, M_2(v) \rangle \qquad (15.2.2)$$

3 A REFORMULATION OF THE ABSTRACT EXISTENCE PRINCIPLE

There is an important necessary condition that must be satisfied by the given vector $\varphi \in B_1{}^*$ in order for it to be compatible with (15.2.2) [or (15.1.1)], and this was not previously mentioned here. Let V_2 be the kernel of the linear operator $M_2: V \to B_2$. Then, we must have

$$\forall v_2 \in V_2, \qquad \langle \varphi, M_1(v_2) \rangle = 0; \qquad (15.3.1)$$

otherwise, it is not possible to solve (15.2.2) since every linear functional ψ will take the zero element of B_2 into 0. See Fig. 67. Let $M_1(V_2)$ denote the image of V_2 on B_1 for M_1 and let the closure be denoted $\overline{M_1(V_2)}$. Further, let $Q = B_1/M(V_2)$ be the factor space. Let $\mathscr{M}_1$ be the linear operator that maps $v \in V$ into the equivalence class $[M_1(v)] \in Q$.

The above abstract existence principle can then be restated as follows:

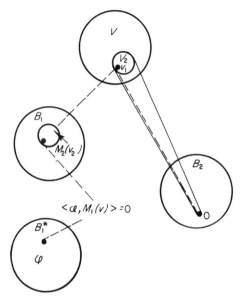

FIG. 67 Compatibility.

Theorem A solution ψ of the functional equation (15.2.2) exists for any function φ satisfying (15.3.1) if and only if

$$\exists K \in R^1 \ni \forall v \in V, \qquad \|\mathscr{M}_1(v)\|_Q \leqslant K \|M_2(v)\|_{B_2}. \qquad (15.3.2)$$

Let $\mathscr{A}$ be the (closed) subspace of $B_2{}^*$ composed of vectors ψ that satisfy the homogeneous problem

$$\forall v \in V, \qquad \langle \psi, M_2(v) \rangle = 0. \qquad (15.3.3)$$

Denoting by $\mathscr{F}$ the Banach factor space $B^*/\mathscr{A}$, for every $\Phi \in Q$ there exists $\psi \in \mathscr{F}$, uniquely determined, such that $\psi \in \Psi$ implies ψ is a solution of (15.2.2). The element Ψ of $\mathscr{F}$ corresponding to $\Phi \in Q$ satisfies

$$\|\Psi\|_{\mathscr{F}} \leqslant K \|\Phi\|_Q. \qquad (15.3.4)$$

The inequality (15.3.4) is said to be the dual of (15.3.2).

4 APPLICATION OF THE REFORMULATED PRINCIPLE TO $\mathscr{L}^p$-WEAK EXISTENCE

We return to considerations of the boundary-value problems of elliptic–parabolic type. Let L be of elliptic–parabolic type and let L^* be its adjoint.

With $u \in C_L$ and $v \in C_{L*}$, from (12.3.1), (12.3.5),

$$\int_A (vLu - uL^*v)\, dA = \int_{\Sigma^3} [ua^{hk}v_{x_h}n_k - va^{hk}u_{x_h}n_k - uvb]\, ds - \int_{\Sigma^1 \cup \Sigma^2} uvb\, ds.$$

$$(15.4.1)$$

If u satisfies the boundary condition

$$u = 0 \text{ a.e. in } \Sigma^2 \cup \Sigma^3, \tag{15.4.2}$$

then (15.4.1) reduces to

$$\int_A [uL^*v - vLu]\, dA = \int_{\Sigma^1} uvb\, ds + \int_{\Sigma^3} va^{hk}u_{x_h}n_k\, ds. \tag{15.4.3}$$

We now let V be the linear manifold of functions $v \in C_{L*}$ such that the right side of (15.4.3) is zero for every $u \in C_L$, and satisfying the boundary condition (15.4.2). Then putting

$$Lu = g \tag{15.4.4}$$

it follows that

$$\forall v \in V, \qquad \int_A vg\, dA = \int_A uL^*v\, dA. \tag{15.4.5}$$

We will say that our boundary-value problem for elliptic–parabolic type with homogeneous data on $\Sigma^2 \cup \Sigma^3$ (i.e., with $f = 0$) admits an $\mathscr{L}^p$-weak solution if for $g \in \mathscr{L}^p(A)$, $p \geqslant 1$, a function $u \in \mathscr{L}^p(A)$ exists satisfying (15.4.5).

From our reformulation of the *abstract existence principle* and the (Riesz) representation of continuous linear functionals on $\mathscr{L}^p$ by $\mathscr{L}^p$ functions, this general condition follows for existence of $\mathscr{L}^p$-weak solutions: An $\mathscr{L}^p$-weak solution of the homogeneous boundary-value problem of elliptic–parabolic type exists for every $g \in \mathscr{L}^p(A)$ such that

$$\forall v_0 \in V \text{ with } L^*(v_0) = 0, \qquad \int_A v_0 g\, dA = 0 \tag{15.4.6}$$

if and only if

$$\exists K \in R^1 \ni \operatorname*{glb}_{v_0 \in V_0} \left(\int_A |v + v_0|^{p/(p-1)}\, dA \right)^{(p-1)/p} \leqslant K \left(\int_A |L^*(v)|^{p/(p-1)}\, dA \right)^{(p-1)/p}$$

$$(15.4.7)$$

where V_0 is the subset of V of all functions which are solutions of the homogeneous adjoint equation $L^*(v) = 0$.

247

When (15.4.7) holds, then an $\mathscr{L}^p$-weak solution satisfies the inequality

$$\operatorname*{glb}_{u_0 \in U_0} \left(\int_A |u+u_0|^p \, dA \right)^{1/p} \leqslant K\left(\int_A |g|^p \, dA \right)^{1/p}, \qquad (15.4.8)$$

where U_0 is the class of $\mathscr{L}^p$ functions satisfying

$$\forall v \in V, \qquad \int_A u_0 \, L^*(v) \, dA = 0. \qquad (15.4.9)$$

The inequality (15.4.8) shows that the solution u depends continuously in $\mathscr{L}^p$-norm on the datum g, modulo eigensolutions of the problem, and the inequality (15.4.7) is indeed true since it is the dual of a priori inequality (12.4.1).

5 UNIQUENESS OF $\mathscr{L}^p$-WEAK SOLUTIONS

We have given a proof of uniqueness of regular solutions in Chapter 13, and the above proves uniqueness of $\mathscr{L}^p$-weak solutions up to an eigensolution of a problem, but to some extent this begs the question of uniqueness for $\mathscr{L}^p$-weak solutions to the Fichera problem. In an interesting adaptation of the Fichera technique, Oleinik has proven this uniqueness, and, incidentally, obtained a formulation for error bounds of weak solutions. We do not give a treatment of these results here, but only indicate the direction of the work.

Where R is a region in E^n, we say for $F: R \to R^1 \wedge \beta \in R^1$, that $F \in C^{(\beta)}(R)$ if F satisfies the following Hölder (or order β Lipschitz) relation in Euclidean norm:

$$\forall x, y \in R, \qquad \exists K \ni |F(x)-F(y)| < K\|x-y\|^\beta.$$

Alternately, where Ω is the set of positive integers, for $N = \{x \,|\, (x \in \Omega) \wedge (x < \beta)\}$, $m = \max N$, $F \in C^{(m)}(R)$ and all directional derivatives of F of order m satisfy a Hölder condition of order $\alpha = \beta - m$.

The smoothness assumptions for coefficients of the elliptic–parabolic operator L were slightly strengthened in the Oleinik work. It is there assumed that there is a region $D \supset \bar{A}$ (where $\bar{A}$ denotes the closure of our region A on which a boundary-value problem is assumed), such that

$$a^{ij} \in C^{(2+\alpha)}(D), \qquad b^i \in C^{(1+\alpha)}(D), \qquad c \in C^{(\alpha)}(D)$$

where $0 < \alpha < 1$. The following theorem is stated:

Theorem Suppose there exist $c_1, c_2 \in R^1$ such that

$$c^* \leqslant c_1 < 0 \qquad \text{and} \qquad 2c - b^i_{x_i} \leqslant 2c_2 < 0,$$

and suppose the set

$$\Sigma^2 \cup \{x \mid (x \in \Sigma^2 \cup \Sigma^3) \wedge (b(x) = 0)\}$$

is closed. If the function $u \in \mathscr{L}^p(A)$, $p = 2$ or $p \geqslant 6$, is such that

$$\int_A L^*(v) u \, dx = 0$$

for any $v \in C^2(\bar{A})$ which vanishes on $\Sigma^1 \cup \Sigma^3$, then $u = 0$ a.e. in A.

The method is a singular perturbation. Oleinik considers the equation

$$L_\varepsilon u = \varepsilon \Delta u + Lu = f, \qquad \varepsilon > 0,$$

in lemmas and an equation

$$\varepsilon \Delta v + L^* v + a \Delta v = \Phi$$

in the proof of the theorem, where it is examined what happens as $\varepsilon \to 0$. Of course, the operators appearing in the left member of both equations are elliptic type. The arguments are difficult and require a detailed knowledge of the Fichera results.

6 PROSPECTUS

We have taken the attitude here that the way in which we should consider the existence of solutions of partial differential equations—especially non-linear equations but linear ones as well—is a matter that is still developing. We now present this, however, only as an attitude, one we would hope would be cultured for the enlightenment it can bring but not accepted as some sort of fact. It may not be so; it may only represent our limited reception of the modern developments. It may be that the body of advanced, sophisticated workers in partial differential equations is ready to accept now one sense of solutions as being of leading importance over all others, though we have thought that the many vigorous contenders for this point of view were unconsciously propagandizing for their particular brand of mathematics.

But what we ask is that the student try to understand the issues and then seek a position that he is satisfied with, not accepting propaganda, especially not if he finds it here, simply on face value or for the cheap rewards it offers to him. This text will have been effective if it has made the student better able to think in this manner and thus better able to direct his studies of partial differential equations.

Toward cementing our perhaps still tenuous understanding of these things, let us enter into a gross discussion of the relative merits of various modern developments. We have seen in Part IV (especially in this chapter) an introduction to the concept of $\mathcal{L}^p$-weak solutions, including a necessary and sufficient condition for existence involving the role of a priori inequalities. But weak solutions can have such very weak smoothness properties as to make their physical value questionable. One may start from a Hamilton principle, clearly involving velocities, and end up with a solution of a partial differential equation which does not have the derivatives necessary to define velocities. Many second-order equations of mathematical physics arise by introducing a potential or streamfunction, and often only the derivatives of these functions have any real physical meaning. We find no significant difficulty, as often emphasized, in deriving the equations of mathematical physics when weak solutions are proposed to be used, but we do see difficulties in physical interpretations of results *in some cases*.

Of course, this was clearly recognized by the early originators of both the theories of weak solutions and solutions in sense of distributions (see our crude description in Chapter 5), and early in these developments there came proposals of techniques for regularization. We have not presented methods of regularization here, but the student can now read about them. They have undergone expansive development, but remain quite difficult indeed. We have seen in the classical pipe flow problem of Chapter 4 that it may be, for those problems with a damping factor involved, one would be better off to accept a weakened sense of uniqueness, amounting to mean-square asymptotic stability with respect to initial data, and thus be enabled to strengthen the sense of existence used. For pipe flow, regular or even analytic solutions can be used, and the difficulties with physical interpretation immediately vanish when it can be shown that the steady state is arrived at with a large (but negative) exponential rate for reasonable values of the dimensionless parameters involved. It may well be that all physical (or even, for example, economic) problems should be studied as time-dependent ones, and time-independent phenomena, when they exist, studied as the time-asymptotic, or steady, state that develops when mean-square asymptotic uniqueness can be proven.

There are rather astounding modern developments we have not been able to develop here; other developments of classical techniques have been extended into modern times and deserve more attention than we have been able to give them. Among the latter are direct methods, a leading proponent being C. B. Morrey of Berkeley, California. The philosophy of this "school" is essentially that since many partial differential equations arise from attempts to minimize functionals (usually quadratic), we might try to prove existence by working with the minimization of such functionals directly. Such an idea could probably be traced back to Euler and surely back to Raleigh in England and Ritz in Germany. However, since the existence of functionals over a class of functions u always make less smoothness demands on u than the partial differential operators that appear in the differential equation, direct methods too represent a sense of weak solutions. It is one, however, that preserves the qualities needed for sound physical interpretations of results. Indeed, the working philosophy of this school would seem to be very oriented to physical applications, but, strangely, its principle proponents have often forthrightly sought occasion to acknowledge that they knew little about applications and modestly denied that they could supply responsible comment on them [33].

In modern times operator theory has assumed great importance. By operator theory is broadly meant a theory of operators (usually linear) on Hilbert and other Banach spaces, which essentially replace the theory of matrices (regarded as linear operators on finite-dimensional linear spaces). Representations of operators are sought, for example, to replace the diagonal form, etc., of matrices with respect to eigenvectors of the matrix. Such attempts go at least back to Hilbert (who, apparently, expressed surprise in later life to hear the term "Hilbert space"). Operators that had such a diagonal form were essentially characterized by Hilbert as "completely continuous," but even the meaning of this term has now been rather completely altered by developments that proceed apace, and it has largely been replaced by the term "compact operators." There are representations of operators in terms of Cauchy or boundary integral formulas (see Chapter 3). The student who wants a quick, understandable orientation to the subject may consult [31] and many other fine sources. Much can be learned about the "inner nature" of operators using the methods of analysis of operator theory, and it is certainly tempting to turn these powerful techniques on the elucidation of the operators appearing in partial differential equations (see Hörmander, *Linear Partial Differential Operators* [20]). But differential operators are unbounded, therefore not continuous, and they are the most difficult of all to be studied. We hope that one who would undertake this difficult area will want to start out with a concrete background like that given in this text even if he finds himself capable of this much more difficult material. It should be noted that operator theory saw its beginnings in the use of differential operators. Laplace, using s as a symbol for a derivative, sought to reduce ordinary differential equations to algebraic ones. Much later, Oliver Heaviside used such symbols to replace partial differential operators and discovered many results only now understood.

Finally, we must not fail to mention that a puzzling flaw in the use of partial differential equations to describe physical phenomena was brought to light by Lewy [27] in 1957 and developed into a remarkable theory by Hörmander [21] in 1961. It was shown that partial differential equations with complex coefficients exist which have no solution, regardless of boundary conditions, even in the weakest possible sense! Hörmander showed that, in a certain Baire-category sense, there are more such equations than there are those that have solutions. Moreover, arbitrarily close to the Laplace equation (in some topology) there is an equation with complex coefficients which has no solution in the weakest possible sense. This introduces an embarrassing stability question about the use of partial differential equations in physics in cases where it makes physical sense for the coefficients to become complex. Some workers have thus contended that all texts on partial differential equations should be rewritten with a deemphasis on boundary-value problems. This view we have resoundingly rejected, feeling that some form of boundary-value problems will be with us for all time in the treatment of scientific problems.

PART I

CHAPTER 1

Section 3

1. The presentation used here of the theory of characteristics, especially the use of our basic question, is modeled after that introduced by Robert Sauer in his text *Anfangswert probleme bei partielln differential gleichungen*, published in 1958 by Springer Verlag.

2. In describing the Lipschitz condition as the criterion to be used for existence and uniqueness of solutions to the initial value problems (1.3.7), (1.3.8), and thus for the existence of two families of characteristics in some neighborhood of a point (x_0, y_0), we have made reference to the basic proof by Picard iteration. Other criterions may serve as well or better according to the reader's tastes in ordinary differential equations. Continuity alone of functions k_j' is not enough to guarantee uniqueness as the trivial example $y' = y^{1/2}$, $y(0) = 0$, quickly shows.

Section 7

1. Concerning the use of inviscid models in large cylindrical tubes, we note that if the Reynold's number $\rho v(r/\mu)$ is large, where r is the radius of the tube and μ is the coefficient of viscous shear stress between fluid layers, then effects of the walls are considered small in the central flow, and the inviscid assumption is considered to be valid. See also, a note below for Chapter 4, Section 7.

CHAPTER 2

Section 5

1. The displacement vibration of a taut string was one of the earliest problems to challenge the metal of mathematical scientists. In fact, a great controversy raged involving Euler, the great Bernoullis, and others over whether a continuum or bead model was appropriate for description of this phenomenon, and this question was finally resolved by a theorem which eventually appeared in a treatise on mechanics by Lagrange, published about 1800, showing that if three or more beads per wave length are used, the two treatments give equivalent results. It is doubtful, of course, whether the dichotomy between continuum and particle models will ever be put to rest entirely; this was but the first of many revelations that these two seemingly divergent views of matter can under some conditions be equivalent and equally useful. Luckily, this early controversy was not too quickly settled, for many ideas and techniques now considered fundamental in analysis were born in the dust of this vigorous quarrel. Also, see Note 1 of Chapter 8, Section 6.

Section 6

1. This method of proving uniqueness can be regarded as an extremely simple "application" of the very sophisticated method of energy integrals introduced by Jean Leray. The energy integral would be the last one appearing in (2.6.1). The method is remarkable in that the proof does not utilize any form of characteristic theory or otherwise reveal the wave character of solutions to the problem. Even the "natural" characteristic boundary segments do not appear. In fact, one has a feeling that this is a method for elliptic or parabolic type adapted to use on hyperbolic type problems.

CHAPTER 3

Section 2

1. It was the dream of Hamilton and at least one generation of mathematicians to produce some suitable hypercomplex number system to play the role of complex variables for the higher dimensional Laplace equation. Any analytic function of a complex variable whose derivative at the origin is not zero is a conformal mapping (harmonic preserving), and this gives the theory of complex variables much of its power. Hamilton's dream was finally laid to rest when it was proven that just four (one nontrivial) harmonic-preserving maps exist in higher-dimensional space. Quarternions now survive as an interesting example of an algebra, not for their interest in analysis.

Section 6

1. It may be noted that the design for equipotential surfaces to give Peirce-type electron beams a prescribed shape have often been devised by solving the Cauchy problem for the Laplace equation, but this always involves the particular case $g = 0$. Also, the flow behind a detached or bowed shock wave was computed by P. R. Garabedian and the author [15] using a special finite-difference procedure for a Cauchy problem for the streamfunction equation related to axially symmetric, time-independent, inviscid flow. Exercise 18 in Chapter 1 shows that this equation is of elliptic type if the flow is subsonic, and the flow is so behind the normal portion of the shock.

CHAPTER 4

Section 6

1. Gevrey was interested in this problem because it reveals under what conditions solutions of the slab or other heat equation problems may be uniquely continued across side (noncharacteristic) boundaries.

Section 7

1. Observation (4.7.9) was epoch-making because prevalent theories utilized inviscid models. Since the fluid must be presumed to slip along the walls in such models, it seemed evident that a square power, or area, law should apply. Poiseuille was a medical doctor interested in blood flow, and it had become fairly clear when he undertook his investigations that the area law was not correct. Others anticipated his work, but his results were by far the most convincing and resulted in changing the thinking of the world in respect to viscosity effects.

Section 9

1. From the result of Exercise 10 it is seen that (4.9.4) is not regular in rectangular coordinates while it is in polar coordinates. The equation (4.9.1) comes out of a general theory of Gaetano Fichera of The University of Rome, and Part IV of this book is devoted to a treatment of this theory. Fichera's theorems there will be seen to give a maximum theorem for regular solutions of the problem (4.9.1), (4.9.2), and this yields uniqueness immediately, of course. We have seen that our (4.9.4) is not a regular solution in the coordinates of (4.9.1), but a theorem of Olga Oleinik, an eminent Russian analyst, covers this case. If in (4.9.1), terms xu_x and yu_y are replaced by $2xu_x$ and $2yu_y$, the solution is regular and it is C^∞. This is the equation presented by Fichera. The author has investigated numerical approximations of solutions to it.

2. The electrophoresis technique mentioned at the end of Chapter 4 is one for separating macromolecules widely used in hospitals and in biological research. In the simplest case

$$Du_{xx} = u_t + Eu_x = du/dx$$

where du/dx is on a curve $dx/dt = E$. Here u is the concentration of a component, D is a diffusion coefficient of the component in a diffusive medium, and E is the electrophoretic mobility of the component being forced through the diffusive medium by an electric field. The whole is an example of forced diffusion.

CHAPTER 5

Section 4
1. The sense of completeness that is used here in "completion by ideal vectors" (and even others only vaguely, or shall we say, only philosophically, related to it), has become an essential way of thinking in mathematics, and even in mathematical logic. It is a sense which clearly originated with the basic work in the last century on the nature of our real number system. Thus it cannot be considered the exclusive property or domain of the field of partial differential equations. It is, in fact, the obvious application of sound modern mathematical conceptions used to release us from a context that is found to be too restrictive.

Section 8
1. For the outline, the student need not be concerned about the meaning of Lebesgue integration, but it will not be discussed here any further. Actually, equivalence classes need to be formed to identify functions which differ by one of measure zero. Convergent sequences of $\mathscr{L}^p$ functions converge to an $\mathscr{L}^p$ function, and this gives the theory of Lebesgue integration much of its power. A lack of knowledge of Lebesgue integration will not be a crutch to the student until Chapter 15, and this material he may read, too, with profit before, or simultaneously with, a study of Lebesgue integration.
2. More than one and a half centuries after Lagrange decided to require solutions of physical problems to be regular, there are those who pause to wonder if this did not mislead us, because the difficulty in proving existence for large classes of boundary-value problems has proven to be overpowering.

Yet in using weak, rather than regular, solutions we may sometimes lose contact with the physical significance of our results. This is particularly true in, shall we say, the more concrete problems of mechanics such as the problem of a vibrating string and the area of fluid dynamics where there is sometimes literally nothing that one can give meaning to that is like velocity (a derivative of a potential or stream function or a derivative of displacement). See also comments on existence in Chapter 5, Section 1.

Section 10
1. Professor Tricomi completed his work in Italy on this equation shortly before World War II and was then cut off from contact with much of the world. He had completed his work quite independently, unaware of the application indicated in the work of Chaplygin. During the war an understanding of controlled flight at transonic speeds seemed to be urgently needed. Then and after there was a great upsurge of interest in Professor Tricomi's work, and he became after the war a celebrity of sorts among those with even modest interests in aerodynamics. Professor Tricomi (who is said to be the only person capable of delivering an address in three languages—all used at the same time) said that he had, of course, heard of these things in Italy but had thought they were something "mysterious," or, as they say in the United States, "classified."

PART II

1. For iterative treatment of the usual integral equation associated with (II.2) the Italians claim priority for Guiseppe Peano (1858–1932) by 2 years and refer to Peano–Picard iteration.
2. The integral equation for u one gets from the initial-value problem for the equation $u_{xy} = f$ must be augmented in our arguments by ones for u_x and u_y since f may depend on these quantities as well as on u. Thus one may regard the integral equation as a system of integral equations or as an integral equation in a vector-valued function with components u, u_x, and u_y. However, the equations for u_x and u_y are gotten as simple derivatives of that for u, so this view is not really justified unless a general theory of contractive maps on a complete metric space is to be used in proving convergence of the iteration involved. We do not do this here because it loses some of the generality and precision for the special problem treated, and we feel that some instructive value would be lost.

CHAPTER 7

Section 2
1. Georg Friedrich Riemann (1826–1866) died young, and great minds have ever since pondered on what direction he would have moved the world of science. Apparently the Riemann function was conceived in order to try to analyze flows of solid materials under large dynamic loads, a subject far ahead of his time.

CHAPTER 8

Section 1
1. It required the *Great Eastern*, five times larger than any ship afloat when built in 1853, to carry the enormous load of cable (and water to keep the gutta-percha insulation from drying out) for the first commercially successful Atlantic cable, completed in 1868.

Section 2
In a landmark paper, "The Theory of the Electric Telegraph," Proceedings of the Royal Society, appearing in May, 1855, Sir William Thompson, later knighted Lord Kelvin for his devoted work on the theory and laying of the Atlantic Cable (see [6]), flatly assumed that $l = 0$, although Faraday (see [11, pp. 508–520]) had almost gleefully reported in 1854 that cable of the same manufacture as Kelvin's cables had considerable inductance when coiled in water, though not in air. Also, one as well-versed as Kelvin would have known of the 1857 paper of Kirchhoff, containing the more general equation $(8.1.7)$[†], on or even before publication. He would not have had time to use it in designing the first Atlantic cable. It was completed, then broken—in a manner that raised a scandal around Kelvin's antagonist, Woodhouse—during the next year (1858) before it could be put into commercial service. However, Lord Kelvin clearly did have the time to use the more general equation in his 1865 and 1866 cables, the ones the *Great Eastern* laid (see [9]). There was not enough induced voltage [see $(8.1.4a)$] in these cables to justify the inclusion of inductance effects because their basic design caused a quick decay of current so that no large rate of current could be maintained. One can anticipate from $(4.5.1)$, $(4.5.2)$, $(4.5.3)$, and $(4.5.6)$ that no matter how sharp a signal

---

† but containing no solution!

might be introduced onto the Kelvin line, in a brief time it would flatten out sickeningly. Though there was no recognizable wave motion, fortunately, at any given position on the line, there was a maximum effect on voltage; and this could be detected as the passage of a signal. Of course, only a "yes" or "no" was necessary to transmit messages by Morse[†] code. The time to maximum effect was found to be proportional to the square of the distance from the point of origin of the signal, and this quickly became known as "Kelvin's law." If $g = 0$, the time to maximum effect will be found to be

$$t_{max} = \tfrac{1}{2}rc x^{2},$$ (8.2.2)

at a distance x from the originating point of the signal. But then the signal spread rate will be

$$x/t_{max} = 2x/rcx^{2} = 2/rcx$$ (8.2.3)

which is quite intolerably slow at long distances x. Capacitance was also large for these cables. This knowledge of the true difficulties our predecessors faced causes us to shift our admiration from their ability to design and manage the massive enterprise required to put the first Atlantic cable into service, to their ability to use such lines, expecially to use them for a commercially profitable communications enterprise. Partly, this now focuses our attention to Kelvin's extremely sensitive marine galvanometer, a mirror galvanometer consisting of small mirrors attached to a small electromagnet hung from a fine thread. The principle of such a device had been demonstrated in England some years before Kelvin used it by Emil Du Bois-Reymond, eminent Berlin physicist (of French Hugenot descent), who later became the accepted fountainhead of modern electrophysiology. This is the older brother, by twelve years, of Paul whom mathematicians know for his work on calculus of variations and for his motivation of Hardy in his work on $O(h^{n})$ type arguments (*infinitärcalcül*). Paul also brought a resounding close to a certain classical era in the mathematical study of transmission lines. Whatever history may regard him to be, Kelvin thought of himself as a physicist, or even an electrician, relying heavily on his remarkable abilities to perform empirical tests and to design highly sophisticated, sensitive equipment. However, the savory flavor of his work would seem to lie in the manifestly evident fact that he lived in a time when the distinctions between mathematicians, physicists, and engineers were just beginning to be made. He and others of his time (and somewhat later) like the flamboyant "Little Giant," Isambard Kingdom Brunel, designer and builder

---

† Samuel F. B. Morse was a consultant to the cable laying company.

of that gargantuan iron ship, the *Great Eastern* [9]; like Delaval, Swedish creator of the supersonic nozzle; like Reynolds, discoverer of the phenomenon of turbulence through his studies of the flow properties found in inadequate plans for a new Parisian water system, created a golden age of engineering, perhaps never to be excelled. All had a remarkable capacity for deep mathematical thought and creativeness, and in the works that made them famous, this fact shines forth in blinding clarity. But these men also had a grasp of reality which was probably quite unknown to many of the preceding generation of mathematicians.

Section 6

1. Michael Idvorsky Pupin, a Serbian emigrant from the small Austrian military frontier hamlet of Idvor, and American inventor, discovered at the turn of the century how to use inductance coils judiciously placed on a line at discrete points to achieve the effect of a large line inductance smeared over the whole line. His motivations harked back to a theorem in Lagrange's analytical mechanics, published a century earlier, in which the old controversy (see note 1 of Chapter 2, Section 5) between continuum and discrete, or "bead," models was laid to rest by stating that if there were at least three beads per wavelength, then the bead model gave the same answer as the continuum model gave. The inductance was a "load," just as the beads were, and Professor[†] Pupin was used to thinking of loads in this analogy from his peasant boyhood training of listening to ground-waves of swords struck into the ground by his playmate-herders and tapped to signal the approach of possible oxen thieves. The inductance loaded line and the bead model reported by Lagrange were mathematically analogous, and this was the germ of the idea. Much more mathematical work went into the design of the coils themselves and much testing of model lines (of foil) was completed in the laboratory before a practical result was achieved, but the end result revolutionized our communications systems and thus all of society.

2. Prior to the turn of the century, New York was becoming overburdened with extremely heavy systems of hanging wire; and telephony, poor in any case, was limited to spans of 25–40 miles. Pupinization of lines allowed them to be quite light, and balance of leakage effects allowed them to be buried.

[†] of Columbia University, New York, where he built the first electrical laboratory. He was both one of the founders of The American Physical Society and The American Mathematical Society.

PART III

CHAPTER 10

Section 6
1. The material appearing in the remaining part of this chapter was taken from the introductory portion of an Institute of Fluid Dynamics and Applied Mathematics Seminar Series delivered by Professors L. E. Payne and H. F. Weinberger at The University of Maryland in 1955.

CHAPTER 11

Section 2
1. The only valid criticism we have ever heard of Maxwell (a Scotsman) was that he was attentive only to work done in the British Isles—in fact, to that of Faraday—and that he thus ignored the fine work of many giants such as Kirchhoff on the continent. As a consequence, his work was widely ignored by scientists on the continent who naturally found it to be quite difficult. We have it from first hand witness (Pupin [37]), however, that Professor Helmholtz at the University of Berlin admired the work of Maxwell, and, from this motivation, instructed his student Hertz to undertake experiments demonstrating the existence of electromagnetic waves according to the predictions of Maxwell. In fact, Professor Pupin in his autobiography [37] says that Helmholtz was disdainful of his colleagues on the continent for not taking up the work of Maxwell more readily.
2. Pupin also felt strongly that either Maxwell or Henry would have been the first to detect electromagnetic waves (as much as fifty years earlier than Hertz) had they not both accepted responsible administrative posts just as they should have undertaken the task. The Henry gap oscilator that Hertz utilized had become a standard physics demonstration unit by the time he used it!

Section 8
1. If one wishes to claim that his intuition verifies that the solution of the wave equation does not depend on values at interior points of the sphere $S(at, P)$ (see Fig. 54), then he must explain why his intuition tells him this in four-dimensional space–time, where it is true, but does not in three-dimensional space–time where it is not.
2. In a jocular mood the late master of our science, Hermann Weyl, who was imbued with the importance of the Huyghens principle to mathematical physics said, "that there was darkness and then there was light would work in

any space, but if God ever expected that he may want to turn the lights off again, it had better have been in a space with a Huyghens principle." This elucidates the sharp "cut-off propagation" that typifies the Huyghens principle. Probably there is no nontrivial fact of experimental science that is so well established as that we do indeed live our lives in a world where the Huyghens principle operates.

PART IV

CHAPTER 14

1. A generation of mathematicians have revered the Hausdorf maximum principle as Zorn's lemma. However, Professor Max Zorn insists that he never stated a lemma in this generality, and the style of attributing this general principle to him was incorrectly started by "the Bourbaki school." We have chosen to respect Professor Zorn's wishes and call this most indispensable mathematical tool by a name which is actually popular in another school of mathematical thought.

CHAPTER 15

Section 6
1. We feel that we are not able to decide once and forever what sense of existence is the best to be accepted by workers in physics, engineering, physiology, or in whatever other fields workers may choose to adapt their ways of thinking to the techniques of partial differential equations. One must look at specific problems in these applied fields, within the context of modern mathematical and scientific (epistomological) frameworks, in order to decide such an issue.
2. There are those who are not ready to accept our weakened concept of uniqueness, and surely we must try to understand their position. They would like to examine all solutions, including those we may wish to call "transients." Jean Leray has proved that all weak solutions of the Navier–Stokes equations become regular in time, a fact that would be expected and one that is well appreciated in the discussion of mean-square asymptotic uniqueness, but presumably it is not given by Leray for that purpose. Many years ago now, Eberhard Hopf [19] derived, by way of a somewhat heuristic argument, an ordinary functional differential equation governing a probability measure of the class of all solutions to the Navier–Stokes equations. The Hopf equation,

much beloved by those who favor it as containing a key to the physical phenomenon technically described as turbulence, has been carefully restudied over the years by many authors (including its originator) and sometimes reworked into slightly different forms. Perhaps such workers will want to study the probability measure of solutions to the Navier–Stokes equations satisfying the wall boundary condition (that the velocity vector vanishes at the wall) rather than simply accept a demonstration of mean-square asymptotic uniqueness even for tube flow.

REFERENCES

1. Apostol, T. M., *Mathematical Analysis*. Addison-Wesley, Reading, Massachusetts, 1957.
2. Bazley, N. W., Lower bounds for eigenvalues with application to the helium atom, *Phys. Rev.* [2] **120** (1960), 144–149.
3. Bryan, C. A., On the convergence of the method of nonlinear simultaneous displacements, *Rend. Circ. Mat. Palermo* [2] **13** (1964).
4. Coddington, E. A., *An Introduction to Ordinary Differential Equations*. Prentice-Hall, Englewood Cliffs, New Jersey, 1964.
5. Courant, R. and Hilbert, D., *Methods of Mathematical Physics*, Vol. II. Wiley (Interscience), New York, 1962.
6. Dibner, B., *The Atlantic Cable*. Ginn (Blaisdell) Boston, Massachusetts, 1964.
7. Diaz, J., Upper and lower bounds for quadratic functionals, *Collect. Math.* **4** (1951), 3–50.
8. Diaz, J., and Weinstein, A., The torsional rigidity and variational methods, *Amer. J. Math.* **70** (1948), 170–116.
9. Dugan, J., *The Great Iron Ship*. Harper, New York, 1953.
10. Faraday, M., *Experimental Researches In Electricity* (collected works), Vol. III. Dover, New York, 1965.
11. Fichera, G., Sul miglioramento delle approssimazioni per defetto degli autovalori I, II (English summary). *Atti Accad. Naz. Lincei Rend. Cl. Sci. Fis. Mat. Natur.* [18] **42** (1967), 138–145, 331–340.
12. Fichera, G., Approximation and estimates for eigenvalues, in *Numerical Solution of Partial Differential Equations* (J. Bramble, ed.), pp. 317–352. Academic Press, New York, 1966.
13. Fichera, G., On a unified theory of boundary value problems for elliptic-parabolic equations of second order, in *Boundary-Value Problems In Differential Equations*. (R. Langer, ed.), Univ. of Wisconsin Press, Madison, 1960.
14. Fichera, G., Alcuni recenti aviluppi della teoria dei problemi al contorno per le equationi alle derivate parziali, *Atti del Conv. Int. Trieste, 1954*; Su un principio di dualita per talun formole di maggiorazioni relative alle equazioni differenziali, *Atti*

264

REFERENCES

Accad. Naz. Lincei Rend. Cl. Sci. Fis. Mat. Natur. [8] **19** (1955), 411–418 (1956); Sulla teoria generale dei problemi al contorno per le equazioni differenziali lineari, *ibid.* [8] **21** (1956), I, 45–55, II, 166–172.

15. Garabedian, P. R., and Lieberstein, H. M., On the numerical calculation of detached bow shock waves in hypersonic flow, *J. Aero. Sci.* **25** (2) (1958), 109–118.
16. Garabedian, P. R., *Partial Differential Equations.* Wiley, New York, 1964.
17. Gevrey, M., Sur les equations aux derivees partielles du type parabolique, *J. Math. Pures Appl.* [6] **9** (1913), 305–471; **10** (1914), 105–148.
18. Gould, S. H., *Variational Methods for Eigenvalue Problems, an introduction to the Weinstein method of intermediate problems,* 2nd ed. Toronto Univ. Press, Toronto, 1966.
19. Hopf, E., A mathematical example displaying features of turbulence, *Comm. Pure Appl. Math.* **1** (1948), 303–322.
20. Hörmander, L., *Linear Partial Differential Operators.* Springer, Berlin, 1963.
21. Hörmander, L., On existence of solutions of partial differential equations, *Symp. Partial Differential Equations and Continuum Mechanics* (R. Langer, ed.), pp. 233–240. Univ. of Wisconsin Press, Madison, 1961.
22. Johnson, R. E., and Kiokemeister, F. L., *Calculus with Analytic Geometry,* 3rd ed. Allyn & Bacon, Boston, 1965.
23. Kamke, E., *Differential Gleichungen Reeler Functions.* Chelsea, New York, 1947.
24. Kellogg, O. D., *Foundations of Potential Theory.* Springer, Berlin, 1929; Dover, New York, 1953.
25. Kirchhoff, G. R., Über die Bewegung der Elektrizität in drahten, *Prgg. Ann.* **100**, *Ges. Abh.* (1857), 131
26. Kolmogorov, A. N., and Fomin, S. V., *Elements of the Theory of Functions and Functional Analysis,* Vol. 1, *Metric and Normed Spaces.* Graylock Press, Baltimore, Maryland, 1957–1961.
27. Lewy, H., An example of a smooth linear partial differential equation without solution, *Ann. Math.* [2] **66** (1957), 155–158.
28. Lieberstein, H. M., *A Course in Numerical Analysis.* Harper, New York, 1958.
29. Lieberstein, H. M., *Mathematical Physiology—Blood Flow and the Electrical Activities of Cells.* American Elsevier, New York, 1972.
30. Lieberstein, H. M., and Mahrous, M. A., A source of large inductance and concentrated moving magnetic fields on axons, *Math. Biosci.* **7** (1970), 41–60.
31. Lorch, E. R., *Spectral Theory,* Oxford Univ. Press, New York, and London, 1962.
32. Moon, P., and Spencer, D. E., *Foundations of Electrodynamics.* Boston Tech. Publ., Boston, 1965.
33. Morrey, C. B., Existence and differentiability theorems for variational problems for multiple integrals, in *Symp. Partial Differential Equations and Continuum Mech.* (R. Langer, ed.), pp. 241–270. Univ. of Wisconsin Press, Madison, 1961. ▪
34. Picard, C. E., Sur l'equation aux derivees partielles qui se recontre dans le theorie de la propagation de l'electricite, *Compt. Rend.* **118** (1894), 16–17.
35. Poiseuille, J. L. M., Recherches experimentales aur le mouvements les liquides dans les tubes de tres petits diametres. Memoir presentes pars divers savants a l'Academie royale des sciences de l'Institute de France **9** (1846), 433–545. Translation by Winslow H. Herschel, Easton, Pennsylvania, 1940, Experimental investigations upon the flow of liquids in tubes of very small diameter.
36. Polya, G., and Szeqo, G., *Isoperimetric Inequalities of Mathematical Physics.* Princeton Univ. Press, Princeton, New Jersey, 1951.

REFERENCES

37. Pupin, M. I., *From Immigrant to Inventor*. Scribners, New York, 1970 (original copyright 1922).

38. Rayleigh, J. W. S., *Theory of Sound*, Vols. I and II. Dover, New York, 1945.

39. Royden, H. L., *Real Variables*. Macmillan, New York, 1963.

40. Varga, R. S., *Matrix Iterative Analysis*. Prentice-Hall, Englewood Cliffs, New Jersey, 1962.

41. Watson, G. N., *A Treatise on the Theory of Bessel Functions*. Cambridge Univ. Press, London and New York, 1944.

42. Webster, A. G., *Partial Differential Equations of Mathematical Physics* (S. Plimpton, ed.). Stechert, 1933.

43. Weinberger, H., Lower bounds for higher eigenvalues by finite difference methods, *Pacific J. Math.* **8** (1958), 339–368; erratum 941.

44. Weinstein, A., An invariant formulation of the maximum-minimum theory of eigenvalues, *J. Math. Mech.* **16** (1966), 213–218.

45. Weinstein, A., some numerical results in intermediate problems for eigenvalues, in *Numerical Solution of Partial Differential Equations*, pp. 167–191. Academic Press, New York, 1966.

46. Young, D., Jr., Iterative methods for solving partial differential equations of elliptic type, *Trans. Amer. Math. Soc.* **76** (1951), 92–111.

Index

A

Absorbed as a scale, 63
Abstract, 74
Accelerated form, 56
Accelerated successive replacements, 56
Accelerated version, 55
Acceleration, 66
Acceleration parameter, 56
Accuracy, 54
Additions, 150, 238
  natural, 85
Additivity, 226
Adiabatic state law, 25
Adjoint, 90, 246
  symmetry, 120, 121, 123
Administrative posts, 261
Aerodynamics, 257
Agreements, 222
Analogous,
  algebraic problem, 45
  mathematically, 260
Analysis, 56, 126, 224, 225
  classical, xiii, 145
  modern, 54, 73, 74, 166, 185
Analysis situs, 88
Analyticity, 99
Anticipations, 76, 77, 97
Apostle, T. M., 142
Applications, xi, 169, 214, 251

Approximating family, 217
Approximation, xi, 37, 53, 56, 126, 216,
  219
  least squares, 118, 220
  method of computing, 54, 118
  methods, 53, 73, 69, 218
  numerical, 12, 19, 119, 215
  polygonal, 20, 21
Area law, 255
Argument, a posteriori, 185
Artificial point, 39
Assumption, 92
Asymptotically unstable, 63
Asymptotic behavior, 61
Augmented matrix, 21, 22
Axial directions, 25
Axially symmetric, 24, 27
  flow, 71, 255
Axiom of choice, 224
Axons, of muscle tissue, 98
  of nerve cells, 98, 141

B

Baire category sense, 252
Balanced line, 137
Banach–Steinhaus theorem, 221
Barriers, natural, 14, 15, 65
Basic problems, 36
Beads per wavelength, 260
Bernoulli law, 26, 27

Bessel function, 70
  of order zero, 127
Binomial theorem, 38, 152
Biological research, 256
Blood flow, 72, 255
Body force, 66
Bound,
  lower, 169
    of eigenvalues, 176
  upper, 107, 169, 224
    of eigenvalues, 176
Boundary,
  closed, 42, 154
  function, 50
    continuous, 82
  natural, 59, 94, 254
  noncharacteristic, 255
  point, 48
  terms, 102, 104, 173
  three sets of, 83
  topological,
Boundary conditions, homogeneous, 43,
  175, 188, 218
Boundary-layer theory, 72
Boundary-value problems, 45, 59, 68,
  82, 83, 86, 92
  bizarre, 71
  deemphasis, 252
  homogeneous, 43, 207
  initial, 69
  two point, 72
Bourbaki school, 262
British telegraph system, 137
Bryan, Charles, xii
Built on an open rectangular base, 102

C

C^∞, 255
Cable, Atlantic, 258
Cantor formulation, 85
Capacitance, 23, 132, 129, 259
  electrostatic, 171
Capacitive current, 24
Capacitor, perfect, 132
Cauchy, 145
  data, 31, 75, 77, 139, 147, 153
    curve, 18
  integral formula, 47, 48, 50, 251
  integral theorem, 47
  problem, xiii, 3, 18, 19, 21, 28, 29, 31,
    32, 36, 44, 56–58, 63, 65, 75, 76,

78–80, 85, 97, 98, 114–117, 128,
    129, 134, 145, 155, 177, 183, 189,
    190, 192, 193, 197, 255
  for the Laplace equation, 255
  on an infinite line, 63
  sequences, 85
Cauchy–Kovalevsky theorem, xiii, 16,
  48, 98, 99, 142, 143, 147, 149, 154,
  155
Cauchy–Riemann equation, 47
Cauchy–Riesz lemma, 203, 210
Cell membrane, 141
Cells, nerve and muscle, 135, 141
Chain rule, 3, 23
Changes of variables, 4
Chaplygin, 95, 257
Characteristic boundary-value problem,
  29, 31, 32, 37
Characteristic cone, 196
Characteristic coordinates, 97, 103
Characteristic direction, 14, 17, 23, 26
Characteristic equation, 14, 21, 25
Characteristic portion of the boundary,
  205
Characteristic problem, xiii, 36, 76, 97,
  101, 102, 103, 110, 116, 121, 124,
  127
Characteristic ray, 77
Characteristic rectangle, 155
Characteristic segment, 14, 16, 18, 20,
  21, 31, 32, 37, 77, 94
Characteristic triangle, 36, 101, 116, 117,
  140
  forward, 76
  retrograde, 76
Characteristics, 3, 9, 14, 15, 17, 18, 23,
  25, 26, 27, 29, 30, 31, 32, 33, 36,
  59, 76, 94, 116, 146, 153, 155, 196,
  199, 253
  angle between, 33, 34, 77
  definition, 12–14, 18
  existence of, 17
  family, 65
    circles, 71
    single, 16
  intersecting, 32
  intersecting segments, 76
  theory of, 12, 18, 19, 21, 24, 28, 29,
    154, 253, 254
  two real families, 15, 16
Choice, optimum, 220

Circuit theory, 131, 132
Circular cylinder tube, 67, 68
Classification, 3, 15, 22, 24, 27, 99
Closed graph theorem, 221, 239, 241
Closure, 42, 245
Cluster of three points, 39
Coalescence, 23
Coddington, E. A., 98
Coefficient matrix, 14, 22, 53, 55, 56, 217
Coefficients, 12
  complex, 252
  continuity conditions, 210
  smoothness assumptions, 248
  successive, 148, 150, 151
Coils,
  at discrete points, 140
  inductance, 260
  loading, 140
  on poles, 140
Communications systems, revolutionized, 260
Compatibility, 18, 19, 21–23, 25, 26, 37, 52, 58, 245, 246
Complete, 85
Completeness, 234, 237, 256
Completion by ideal vectors, 85, 93, 256
Complex conjugates, 14, 22, 26
Complex function theory, 49
Complex numbers, 142
Compliance, 67
Compressibility, 67
Compressible flow, 22, 24
Compression wave, 23, 67
Computability, 110
Computation, 38
  impractical, 218
  procedure, 118
Concentration, 63
Concrete problem, 257
Conditions,
  Hölder, 248
  necessary, 243, 245
  necessary and sufficient, 88, 241, 250
  specified, 78
Conductance, 132
Conducting fluid, 141
Conducting medium, ambient, 132
Conducting surface, 171
Conjugate, 244
Conoid, forward, 178

Conormal, 205
  derivative, 205
Conservation,
  of energy, 26
  of mass, 22
    generalized, 134
Content, 117, 134
Continuations,
  analytic, 16
  of continuous linear functionals, 214
Continuity, 5, 22, 24, 41, 47, 52, 59, 64, 98, 108, 118, 202, 232, 236, 237, 239, 242, 251, 253
  complete, 251
  condition, 42
  equation, 22, 26
  of linear operators, 233
Continuous,
  dependence, 74–76, 82, 84, 110, 118, 248
  differentiability, 6
  extension, 18
  four equivalent conditions for continuous operators, 231
  $N–N'$ continuous, 232
Continuously dependent on data, 73
Continuum model, 22, 36
Contractive maps, 114
Contradiction, 7, 226
Convergence, 65, 98, 105, 117, 148
  absolute, 142
  necessary and sufficient condition for, 55
  point for point, 86
  uniform, 84, 105, 108
Convergent series, 108
Converging–diverging nozzle, 77
Coordinates, 97, 147, 181, 255
  polar, 255
  rectangular, 186, 255
  spherical, 186, 197
Correction matrix, 56
Cosine terms, 40
Couples, ordered, 223
Courant, R., 169
Cramer's rule, 19
Curl, 25
Current, 131
Curvature, 116
Curve,
  analytic, 146, 147

bell-shaped, 134
  flattened, 134
  simple closed, 120
  squashed down, 134
Curve segment, 115
Cylinder, infinitely long, 170
Cylindrical tubes, small, 27

D

D'Alembert solution, 3, 6, 21, 29, 36, 56, 99, 145, 155, 189, 190, 191, 196
  generalization, 186
Damped, quickly, 62
Damping, 69, 93, 136
Data, 29, 31, 32, 46, 57, 61, 63, 65, 74, 83, 94
  analytic, xiii
  on a circle, 57
  continuous, 85
  convergent power series, 48
  derivative, 59
  functions, 79, 84, 94, 99, 126
  homogeneous, 7, 247
  initial, 62, 63, 148, 194
  nonanalytic, 155
  primitive, 33, 34, 42, 59, 76, 77, 94
  side, 38, 62
  smooth, 16, 76
  on the x-axis, 58
Data curve segment, 116
Data segment, 57, 147
  Cauchy, 148
Decaying exponential, 71, 134
Definition, 14, 25
Degenerate case, 72
Delaval, 260
Delta function, 86
De Moivre expression, 51
Density, 22, 27, 66, 67
Denumerable set, 83
Derivative, 257
  on characteristics, 19
  composite, 136
  continuous, 64, 68, 87
  directional, 248
  of displacement, 256
  limits of sequence of, 108
  normal, 52, 60, 76, 164, 177, 205
  of pressure, 66
Derivatives, successive, 49

Determinants, 19, 169
Devices, 22
  control, 169
Diaz, J. B., 174
Dichotomy, 254
Differentiability, 119
Differential, 25
Differential forms, exterior, 87
Differential of velocity, 67
Differentially defined, 25
Diffuse propagation, 136
Diffusion, 136, 137
  coefficient, 63, 256
  forced, 256
  problem, 63, 64
Diffusive spread, 133, 137
Dimensionless parameter, 71
Dimensions,
  higher, 54, 87, 254
  higher-dimensional problems, 58, 78, 146
  two, 54
Dipole,
  charge, 163, 164
  current, 164
  distribution of, 164
  moment, 164
Dirac, 86
Direction cosines, 83, 87, 203
Dirichlet integral, 170, 171, 172
Dirichlet principle, 173, 174
Dirichlet problem, xiii, 42, 43, 46, 48, 50, 51, 52, 53, 57, 58, 69, 71, 76, 78, 83, 84, 91, 154, 159, 160, 161, 166, 167, 168, 169, 170, 185
  for heat equation, 71
  for Laplace equation, 70
  for parabolic type, 71
  for wave equation, 42
Discontinuity, saltus, 146
Discriminator function, 83
Disk, circular, 190
Displacement, 79
Distortion-free transmission, 139
Distortion term, 140
Distributions, 93
  charge, 163
  density, 163
  mass, 163
  theory of, 85, 86

Disturbance,
  point, 194
  spot, 194, 195, 196
Divergence theorem, xiv, 7, 52, 69, 79, 83, 87, 88, 159, 160, 164
Domain, 14, 49
  of dependence, 8
Drag force, 66, 169
Du Bois-Reymond, Emil, 259
Du Bois-Reymond, Paul, 139, 259
Dynamical behavior, 169
Dynamics,
  loads, 258
  problems, 164

E

Economic problems, 250
Eigenfunction, 176
Eigensolution, 248
Eigenvalues, 39, 175, 176, 204, 213
  bounds, xiii
  first positive, 69
  of the Green operator, 169
  of the Laplace equation with respect to the Dirichlet problem, 169
  problems, 169
Eigenvectors, 213, 251
  unit, 204
Elasticity, 169
Electrical analogues, 40
Electrical imbalance, 140
Electric fields, 133, 137
  infinite, 171
Electricians, 259
Electrodynamics, 133, 169, 178
Electromagnetic waves, 261
Electron beams, Peirce-type, 255
Electron, free, 141
Electrophoresis, 72, 137, 256
Electrophoretic mobility, 137, 256
Electrophysiology, 141, 259
Electrostatics, 133
Electrotonic spread, 135
Elements,
  basis, 204
  maximal, 224, 230
  zero, 238, 245
Elimination, 217
Ellipse, 46
Embedded, 85
Empirical tests, 135, 259

Endpoints, 54
Engineering,
  course for students of, xii, 262
  golden age of, 260
Engineers, 75, 85, 176, 259
Enthalpy, 25
Entropy, 24–26
Equations,
  analytic, xiii
  differential, altered, 151
  differential, ordinary, 151, 152, 153
  higher order, 8, 10
  homogeneous, 7, 9, 109, 210, 218
  homogeneous differential, 174, 187, 191
  L-regular, 207
  of least squares, 217
  linear, 8, 9, 93, 95, 113, 121, 199
  linear homogeneous, xiii, 9
  linear homogeneous differential, 86
  of motion, 66, 67
  nonhomogeneous, 102, 139, 207
  nonhomogeneous differential, 86
  nonlinear, 18, 77, 93, 97, 119, 141, 249
  normal, 53, 55
  quasi-linear, 9, 17
  second-order, 76, 78
  second-order linear in n-variables, 11
  type classification, 12
Equivalence, 136, 244
Equivalence classes, 238, 244, 245, 256
Error bound, 54, 84, 110, 118, 214–216, 218
  computable, 55, 218, 219
  L^p, 219
  in $\mathscr{L}^p$-norm, 220
  in maximum norm, 216
  of weak solutions, 248
Error function, 64, 218
  L^2-norm, 219
Error vector, 38
Errors, effects of, 55
Estimates,
  a priori, xiii, 214, 215
  L^p-norm, 210
  maximum norm, 210
Euler, 251, 254
Euler equations, 22, 24, 27
Evolution of mathematical thoughts, 74
Exact differential, 25

Existence, xi, xiii, xiv, 12, 17, 21, 36, 37, 45, 52, 54, 63, 68, 73, 84, 93, 97, 98, 101, 102, 110, 114, 117, 118, 121, 124, 125, 128, 146, 150, 168, 185, 199, 234, 240, 241, 245, 249, 250, 251, 253
  classical, xiv, 114
  concrete extension of meaning, 85
  extension of meaning, 74
  for large classes of boundary-value problems, 74
  of L^p-weak solutions, 214, 246, 247
  of regular solutions, 75
  sense of, 5, 262
Existence and uniqueness, 116
  exploration of ideas related to, 73
Existence principle, abstract, xiv, 199, 214, 234, 239, 241, 245, 247
  first form of, 240
  reformulation, 245, 246
Expectations, 73, 75
Exponential,
  function, 110
  rate, 71
    attenuation, 138, 139
  series, 107, 110
Exponentially amplified, 63
Extension, 230, 242
  of the theory of analytic functions of a complex variable, 48
Exterior problem, 58, 78, 175
Exterior sea, 132
External work, 25
Extrema, 54

F

Faraday, 258, 261
Fichera, G., xi, 17, 71, 83, 84, 95, 176, 199, 200, 234, 239, 248, 249, 255
Finite differences, 39
  direct replacement, 39
  methods, 55, 56
  procedure, 19
  replacement, 118
Finiteness, 114
First-order system, 39, 47
  2×2, 24
Fit, 53
Five-point analogue, 55, 56
Five-point cluster, 55
Flow equations, 22

Flow fields, 169
Flow,
  inviscid, 22, 24, 28, 72, 77, 255
  of materials, xiii, 256
Fluid dynamics, 169, 257
Force-driving, 67
  electrostatic, 163
  gravitational, 163
Forcing function, 102
Form, 40
  bilinear, 170, 172
  sublinear, 230
Formal manipulation, 40
Foundations questions, 88
Fourier problem, 64
  series, 40, 62
Fourth power law, 68
Frankel system of set axioms, 224
Free term, 102, 104
Frequency, independent of, 139
Fubini theorem, 87, 89, 159, 160
Functional analysis, xii, xiii, 73, 123, 176, 199, 214, 221, 230
Functional equation, 240, 245, 246
Functional value, 3, 4, 22, 23, 25, 37, 49, 240
Functionals,
  continuous linear, 123, 236, 240, 242, 243, 245, 247
    extension of, 225
    space of, 221
  linear, 170, 225, 228, 236, 237, 242, 245
  minimize, 251
  ordering on, 228
  positive semidefinite, 171, 172
  quadratic, xiii, 172
  subadditive, 239
  sublinear, 225, 236
Functions, 231
  additive, 225
  admissible, 219
  analytic, 47, 49, 54, 57, 68, 127, 143, 145, 146, 147, 148, 149, 150, 254
  approximating, 54, 217
  approximation, 216
  arbitrary, 4
  with compact support, 91
  complex, 145
  complex valued, 47
  composite, 3, 4, 5, 12, 23

dense in $\mathcal{L}^p$, 92
differentiable, 13, 47
entire, 127
equations, 34, 35
harmonic, 54, 161, 164, 165, 167
homogeneous, 225
initial, 64, 79, 86, 101, 110
integrable, 4, 6
integrable in the Lebesque sense, 91
iterations, 110
kernel, 98
$\mathcal{L}^p$, 91, 247, 248
convergent sequences of, 256
linear, 54, 55, 98, 109, 123, 124, 165, 170, 226
quadratic, 53, 216
real, 49
regular, 199, 218
resolvent, 31, 98, 166, 169
retarded, 180
spaces, 123, 221
stress, 170
subadditive, 225
subdominant, 226
uniquely defined, 115

G

Galvanometer, marine, 259
Gap oscillator, 261
Garabedian, P. R., 150, 255
Gas dynamics, 95
Gas flows, 22
Gauss mean-value theorem, 166
finite difference version, 165
Gauss theorem, 87
General cluster, 19
Generalized functions, theory of, 86
Generalized sense, 86
Geometric series, 35, 144
Gevrey, 65, 255
Goursat problem, 34, 77
Great Eastern, 258
Greatest lower bound, 70
Green's function, xiii, 166, 167, 168, 169
argument, 177
of the second kind, 169
Green's identity, 87, 89, 123, 184, 203
first, 160, 172, 174, 177
generalized, 89, 90, 92, 119, 199, 204, 205

second, 91, 160, 161, 167, 168, 180, 187, 192, 197
third, 160, 161, 162, 164, 165, 166, 185
Green's operator, 168
Green's region, 52, 79, 83, 88, 91, 160, 204
Green's theorem, 173, 174
Ground-waves, 260
Gun barrels, 22

H

Hadamard, xiii, 74, 75, 84, 195
sense of, 73, 76
Hahn–Banach theorem, 214, 221, 222, 224, 230, 236, 242
for normed spaces, 235
Half infinite x-axis, 65
Half-plane, 57
lower, 94
upper, 58, 64
Hamilton principle, 92, 250
Hardy, G. H., 259
Harmonic polynomials, high order, 54
Harmonic preserving, mapping, 58, 254
inversion at a circle, 58
Heat conduction, 63, 72
Heat equation, 59, 62, 63, 74, 82, 85
modified, 84
Heat exchange, 25
Heat flux, 61
Heavifield line, 133, 137, 139–141
Heaviside, O., 137, 251
symbolizing of, 138
proponent *Ernst J. Berg,* 139
Helmholtz, H. L. F. van, 261
Henry, J., 261
Hertz, H. R., 261
High frequencies, 24
Hilbert spaces, 123
theory, 85, 176
HMP, 228, 230
as an axiom, 224
Hodgkin, A. S., 141
Hodograph,
method, 95
plane, 95
Homogeneity, 226
Homogeneous case, 10
Homogeneous problem, 246

Homogeneous wave equation in E^2, 27, 29, 33, 36, 40
Hopf, Eberhard, 262
Hörmander, L., 252
Huxley, A. F., 141
Huyghens' principle, xiii, 189, 190, 193, 194, 261, 262
  in two- or three-dimensional space-time, 195
Hydrodynamics, 171
Hyperbola, 46
Hyperbolicity, 97
Hypercomplex number system, 254

I

Idealization, 67
Identity, 89, 90, 120
  additive, 238
Image, 50, 242, 244, 245
Imaginary parts, 47, 48, 49, 50, 51
Impact loadings, xiii
Implications, cycle of, 232, 233
Impulse problem, 74, 85
Impulses, 86, 136, 137
  form, 134
Incompressible flow, 171
Incompressible fluid, 66, 67
Incompressibility, 27, 67
Indicated sums, 11
Indices, repeated, 11, 82
Inductance, 23, 24, 72, 131, 132, 133, 139
  large, 137, 139, 141, 260
Inductance–capacitance line, 24
Inductance lines, large, 140
Inductance loads, 141
Inductances, continuous distribution of, 141
Induction, 105, 106, 107, 109, 145
  assumption, 106
Inductively defined, 105
Inductor, perfect, 131
Inequality,
  a priori, 54, 185, 199, 201, 214
    in L^p-norm, 208, 209
    in L^∞-norm, 210
  Cauchy, Buniakovski, Schwarz, 172, 175, 203
  dual, 246, 248
  Schwarz–Hölder, 203, 208, 212
  strong, 55

symmetric form, 105
  weak, 55
Inertial force, 22, 66
Inertia, polar moment of, 174
Infimum, 227
Infinite concentration at a point, 63, 86
Infinite line, 63
  doubly, 134
Infinite rods, 63
Infinite string, 195
Infinite strip, 65
Infinitely smooth, 16
Infinitesimal, 145
Inflection point, 135
Initial conditions, 69, 150
Initial iterate, 101
Initial line, 38, 59
Initial segment, 18
Initial temperature, 61
Initial value, 61, 147
Initial value curve, analytic, xiii
Initial value problems, 15, 29, 36, 57, 99, 111, 112, 133, 141, 146, 253, 257
  analytic, 98
  for an ordinary differential equation, 97, 98
Initial vector, 55
Initial velocity, 69
  effect of, 69
Inner normal, 83
Inner product, 245
Instability, 43, 45, 74
  negative times, 62
Instantaneous increment, 67
Institute of Fluid Dynamics and Applied Dynamics, 261
Insulated case, 61
Insulation, gutta-percha, 258
Integers, positive, 142
Integrability, Lebesque, 244
Integral,
  area, 87, 159
  complex, 57
  energy, 254
  infinite, 64
  iterated, 87, 106, 159
  line, 120, 129
  singular, 48
  surface, 87, 160, 194
  volume, 87, 159, 160, 177, 194

Integral equation, 97, 102, 103, 104, 108, 109, 114, 116, 117, 257
Integral identities, basic, of mathematical physics, 92
Integrals with interchangeable limits, 108
Integrand, 41
Integrating factor, 133, 138
Integration,
    Lebesque, 256
    pattern, 104, 116, 122
    theory, 221
Interior point, 48
Interior problem, 58, 78, 174
Internal energy, 25
Interpretations, 153, 162
Intuition, 261
Inverse point, 50
Inversion at a circle, 58
Inversion at a sphere, 58
Ion, 141
ir drop, 24
Irrotational, 25
Isentropic case, 26
Isentropic equation, 25
Isolated points, four, 83
Isomorphism, 245
Iteration, 54, 55, 68, 97, 104
Iterative method, 54, 55, 118, 217
Iterative techniques, linear, 55
*i*th equation, 56

J

Jacobian, 18, 147

K

Kellogg, O. D., 87, 89, 160, 169
Kelvin, 133, 134, 135, 258, 259
Kelvin Law, 135, 259
Kelvin transformation, 58
Kelvin's cables, 134
Kelvin's principle, Sir William Thompson's, 173
Kernel, 245
    singular, 166
Kidney mechanisms, 72
Kirchhoff, 132, 133, 164, 178, 184, 185, 258, 261
    formula, 177
    identity, 178, 183, 192, 197
Kovalevski, Sonja, 145

L

$\mathcal{L}^2$ norm, 54
L^p-norm, 218, 221
$L^p(A)$, 244, 245
$\mathcal{L}^p(A)$, 244, 245
$\mathcal{L}^p$-weak sense, 86
L–C line, 136, 137
Lagrange, 74, 90, 92, 138, 254, 256, 260
Lagrange, problem, 22
Laplace, 251
Laplace equation, 16, 42, 44, 45, 46, 47, 49, 51, 52, 53, 54, 56, 75, 94, 95, 98, 118, 145, 159, 163, 164, 166, 167, 171, 177, 200
    arbitrarily close to, 252
    in higher-dimensional space, 157, 254
    modified, 83, 84
    nonhomogeneous, 91, 164
Laplace operator, 161, 177, 205, 210
Large cylindrical tubes, 22
Large gradient of pressure, 23
Largest circle, interior, 72
Layers of fluid, 67
Leakage, 24, 139
Leakage conductance, 23, 132, 134, 140, 141
    nonlinear, 141
Leakage effects, balance of, 260
Leakage factor, exponential, 134
Leakage resistance, 134
Leaky line, 132
Least-squares, 53, 54, 55, 56
Lebesque integration, xii, xiv, 221
Lemma, 106
Length ratio to other side of rectangle, irrational, 43
    rational, 43
Length ratio,
    to diameter of cell, 63
Leray, J., 254, 262
Lewy, H., 252
Lignes, les-pupinize, 141
Limit function, 75, 108
Limit of solutions, 64
Limiting position, 72
    of a shock, 146
Line, 63
    entire straight, 154
Linear,
    algebra, 19

algebraic system, 55
case, 15, 17
independence, 53, 217
problems, 7, 38, 82, 241
system, 55, 56
Line current, 23
Line leakage, 24
Lines,
buried, 140
overhead air, 140
Lines of values, successive, 38
Lipschitz condition, 15, 18, 97, 102, 106, 108, 109, 110, 111, 112, 113, 114, 117, 124, 252
Local linearizations, 28
Locus, 135
Logarithmic horn, 79

M

Macromolecules, 256
Magnetic fields, 133, 141
Majorant problem, first, 150
Majorize, 108
Manifold,
closed linear, 239
linear, 223, 225, 228, 229, 238, 240, 243, 247
Mapping, 239
canonical, 244
closed, 239
conformal, 254
contractive, 257
linear, 239
N_1–N_2 continuous, 239
Marching scheme, 39
linear, 55
Marcus, Bernard, xii
Marine underground cables, 23, 132, 135, 141
Mass-conservation condition, 63
Matrix,
diagonal, 55, 204
diagonal matrix of eigenvalues, 38
lower diagonal matrix, 55
positive definite, 54, 55, 56, 82
positive semidefinite, 82, 213
real symmetric, 38, 204
similar, 204
symmetric positive-definite, 217
square, 10
upper diagonal, 55

Matrix of coefficients of a linear second-order partial differential operator, 82
Maximum, 202, 228
effect, 259
positive, 213
theorem, 255
Maximum principle, 54, 61, 82, 114, 118, 165, 166, 172, 174, 209
combined, 215, 216
first, 207, 210
Hausdorff, 224, 262
second, 210
Maxwell, 261
Maxwell field equation, 139, 177
Mean square, 40, 71
Mean-value, 165
form, 185
theorem, 111, 162, 163
Measure, 117
positive, 203
Measure zero, 83, 204, 244, 256
Medical doctor, 255
Medium,
homogeneous, 170
isotropic, 170
Melting point, 61
Membrane, infinite, 196
Mesh size, 37
Method of characteristics, 18, 19, 20, 28, 37, 39, 118
Method of descent, xiii, 186, 190, Hadamard, 189, 195
Method of energy integrals, 254
Method of limits, 94
Method of Stöss and Duhamel, 187
Methods, direct, 250
Minimization, 53, 54
L^p-norm, 218
Minimum, 53, 217
negative, 213
Minimum–maximum principle, 165
Minimum principle, 54, 165, 172, 174
Mixed problem, 32, 33, 36, 37, 52, 76, 78, 84, 159, 160
Cauchy and slab, 65
Dirichlet–Neumann, 87
Mixed type, 22
Mode, xii, 1, 2, 3, 4, 5
Model,
continuum and discrete, 260

continuum or bead, 254
  inviscid, 253, 255
Model static versus dynamic, 185
Modern setting, 84
Momentum, 22, 24
Monodromy theorems, 145
Morrey, C. B., 250
Multiplication by a scalar, natural, 85
Multiplications, 150, 238
Myelin, 141

N

Natural boundaries, 30
Navier–Stokes equation, 262
Neighborhood, 47, 48, 57, 143, 145, 146,
  147, 148, 149, 154, 155, 202, 253
  of data line, 98
  of infinity, 58
  spherical, 160
Neumann condition, 174
Neumann problem, 51, 52, 78, 84, 159,
  160, 161, 169
  interior, 174
Neurons, 141
Newtons, law, 22
Nodes of Ranvier, 141
Nomenclature, 8
Noncharacteristic, 18, 21, 32, 33, 65,
  76, 83, 146, 147, 154
  nonintersecting, 33
  ray, 77
  segment, 77
  two, 34
Nonexistence, 79
Nonhomogeneous problem, 42
Nonlinear,
  case, 17
  equations, 22, 24
  principle part, 18, 28
  problem, 109
  theorem, 141
Nonlinearity, 17, 18
  nature of, 24
Nonnormal type, 22
Norm, 109, 230, 231, 235, 240
  Banach, 234
  maximum, 54, 110, 217, 219
  natural, 85
  preservation of, 236
  properties of, 239

spectral, 55, 56
  supremum, 219
Notes, xiv
Not uniquely determined, 18
Nowhere tangent to a characteristic, 18
Numerical analysis, 118
Numerical methods, 169
Numerical quadrature replacements, 118

O

$O(h)$ terms, 20
Oleinik, Olga, 200, 248, 249, 255
Operators, 82, 83, 97, 119, 120, 123,
  126, 170
  compact, 251
  continuous, 232
  continuous linear, 230, 234
  first-order ordinary differential, 133
  first-order partial differential, 136
  linear, 231, 234, 240, 245
  linear second-order partial differential,
    205
  not continuous, 232
  partial differential, 251
  second-order linear, 119
  theory, 251
  unbounded, 251
Ordering, 229
  complete, 224
  partial, 224, 228
Order of taking limits and integrals, 75
Ordinary differential equation, 25, 98,
  110, 131, 141, 146, 253
  first-order, 15
Ordinary functional differential equation,
  262
Orthogonal, 243
Outline, xiii, 1, 145, 159
Overrelaxation, 56

P

Parabola, 135
Parameter, 13, 25, 133
  heavification, 138
  lumped, 133
Partial differential equation, second-
    order, 8
  theories, 72
Partial differential operators, 17
  general, 72
  general second-order linear, 89

Partition, 19
Parts, modular, xi
Payne, L. E., 261
Peano, Guiseppe, 257
Peano-Picard iteration, 257
Performance characteristics, 133, 136
Periodic solutions, 72
Permittivity, 177
Perturbations, 79
  analytic, 57, 79
  regular, 189
  singular, 249
Phenomena, 44, 136, 254, 263
Physical interpretation, 5, 60, 250, 251
Physical meaning, 92
Physically meaningful, 63, 72, 74
Physicists, 75, 85, 86, 259
  modern, usage of, 168
Physics, 169, 194, 262
  course for students, xii
  mathematical, 250, 261
Physiologists, 135
Physiology, 262
Picard, 97, 139
Picard iteration, xii, 101, 117, 126, 140, 253
Picone, M., 169
Pipe flow, classical, 250
Piston drive, 22
Plane,
  complex, 47, 57, 80
  entire, 57, 127
Point,
  discrete, 260
  fixed, 161
  N-limit, 238
  variable, 161
Poiseuille, 68, 255
  flow, 66
Poisson integral, 50, 51
  equation, 91, 164, 166, 171, 185, 200
  kernel, 50
  term, 178
Polar form, 51
Polarization, 171, 174
Polynomial, 22, 106, 150, 151
  Bernstein, 84
  harmonic, 53
Positive constant, 43, 70, 108
Positive terms, 35, 107
Potential, 163, 183

function, 163, 257
retarded, 177, 178, 183
scalar, 177
surface dipole, 164
surface mass or charge, 164
theory, xiii, 47, 49, 58, 87, 159
volume, 164
Power series, 99, 145, 146, 148, 155
  extensions, 48
  in more than one variable, 148
  seven multiple, 143
  in seven variables, 148
  theorem, 145
Preece, William, 137
Preliminaries, 201, 221
Prerequisites, xii
Pressure,
  force, 66
  gradients, 27, 66, 67
    local, 67
    period of, 71
  relief, 67
Primitive, 4
Principle of superposition, 9
Principle part, 9, 17, 18, 137
  with constant coefficients, 97
  linear, 97
Probability measure, 262
Procedures,
  computational, xi, 110
  finite difference, 255
Propagation, 133, 137, 195
  rate, 193
  "sharp cut-off," 195, 262
  "sharp turn-on," 195
Prospectus, 93, 249
Prototype, 10
Pupin, M. I., 140, 260, 261
Pupinization, 260
Pupinized lines, 141
Pupinizierte linien, 141
Pupin–Lagrange smearing, 141

Q

Quadratic form, 17, 93
  positive-definite, 169
  semidefinite, 203
Qualitative properties, 29
Quarternions, 254
Quasi-linear case, 10

Quasi-linear second-order equation, 27
Quasi-linear system, 12

R

Radial direction, 25
Radial variable, 27
Radiofrequency, 24
Radius of convergence, 16
Raleigh–Ritz principle, 69, 176
Rank, 22
Rarefaction, 67
Rate, 139
  of spread, 259
Realization, 140
Real parts, 47, 48, 49, 50, 51
Rectangle, 37, 40, 42, 43, 44, 51, 52, 59, 87, 88, 91, 114, 124
  closed, 102, 103, 105, 113, 116
  open, 102, 103, 105, 114, 115, 120
Rectangular prisms, 88, 159
Rectifiable curve, simple closed, 47
Recursion, 104
Redefinition of G and T, 112
Reference value, 6
Reflected point, 50
Region, 5, 7, 9, 22, 42, 46, 49, 53, 57, 58, 87, 89, 97, 114, 160
  bounded, 102, 112, 114, 117
  exterior, 171, 173
  infinite, 8, 102, 112, 171
  interior, 171
  odd-shaped, 79
  open, 114
  simply connected, 172
  unbounded region, 110, 124
Region,
  of convergence, 144
  of determination, 8, 9, 14, 15, 29, 31, 32, 94, 98
  of influence, 8
  of parabolicity, 72
Regularity, 92
Regularization, 93, 250
Regular perturbations, 24
Relation, 125, 223
  equivalence, 238
Replacements, accelerated successive, 217
Replicas, 136, 137
Representation, 184
  a posteriori, 125
  a priori, 125, 178, 185, 186, 187

integral, xi, 51, 98, 118, 119, 123, 124, 125, 128, 145, 161, 168, 169
  nonsingular integral, xiii
  Riesz, 244, 245, 247
Representations, of operators, 251
Representation theorem, 125, 167
  a priori, 177
Representative equations, 55
Resistance, 23, 131, 132, 139
Resistance–capacitance lines, pure, 64
Resistor, perfect, 131
Restatement, 148
Resumé, 142
Reynold, Osborne, 260
Reynold's number, 67, 69, 253
RF circuits, 24
Riemann, G. F., 258
  function, 120–124, 126, 128, 130, 166, 258
  integral, 108
  method, xiii, 31, 98, 118, 140
Rigid walls, 27, 66, 67
Rod, 61
Rotational flow, 24, 26
Round-off, 38
Roydon, H. L., 222

S

Saint-Venant theory, 170
"Satisfies" a differential equation, 85
Sauer, Robert, 253
Schematic, of abstract existence, 241
Schwarz reflection principle, 49, 57
Self-adjoint, 123, 126
  equation, 138
  operator, 119
Semantics, 222
Semicircular arc, 94
Semicubical parabola, 16, 94
Separability, 225
Separation of variables, 40, 42, 62, 140
Sequence(s),
  of bell-shaped functions, 64
  converging N-wise, 233
  of data, 75
  double, 142
  N-Cauchy, 233
  of problems, 64
  two, 98
Series,
  dominant, 152

double, 142
multiple, 142
  absolutely convergent, 143
single, 142
Set,
  closed, 238
  discrete, 141
  L^p, 245
  of limit points, 46
  open, connected, 7, 46, 87, 203
  permanent, 195, 196
  simply connected, 87
Shock, 23
  detached, 255
  normal portion of, 255
  weak, 147
Shunt, 132
Simplex, 88
Sine series, 39
Sine waves, 40
Singularity, 14, 48
Singular perturbations, 24
Slab problem, 59, 60, 63, 64, 84, 87
Slab-type problem, 69
Solution point, four relative positions of, 117
Solutions,
  energy-weak, 93
  fundamental, 161, 167
  generalized, 93
  in coordinates, 5
  infinitude of nontrivial, 43
  $\mathcal{L}^p$-weak, xi, xiv, 5, 84, 91, 92, 93, 199, 200, 225, 240, 247, 248, 250
  multiplicity of, 74
  no, 252
  nontrivial, 43
  nontrivial separable, 43
  not regular, 255
  not uniquely determined, 18
  power series solution, 48
  regular, xi, 7, 8, 40, 51, 56, 57, 59, 60, 64, 65, 68, 74, 84, 86, 93, 159, 160, 175, 214, 216, 219, 248, 250, 255, 256, 257
  regularized, 93
  regular sequence of, 75
    limit of, 85
  residual, 195
  separable, 45

survival value of, 74
unique, 6, 7, 9, 13, 14, 21, 33, 37, 73, 77, 79, 112, 125 145, 147, 216, 217
weak, 84, 93, 241, 250, 251, 257, 262
Sonic flow, 27
Sonic line, 72, 77
Spaces,
  adjoint, 123, 221, 234, 235, 240, 243
  Banach, 123, 233, 235, 239, 240
  complete metric, 257
  conjugate, 234
  dual, 234, 235
  factor, 238, 245
  Hilbert, 251
  linear, 222, 223, 230, 231, 238, 245
  $\mathcal{L}^p$, 225, 236
  N-complete, 233, 239
  normed, 230
  normed linear, 233, 234, 235, 237, 238, 239
  normed linear vector, 233
  real linear, 225
  second conjugate, 244
  separable, 237
Space–time,
  even-dimensional, 197
  four-dimensional, 189, 193, 195, 197, 261
  odd-dimensional, 196, 197
  three-dimensional, 189, 191, 192, 261
  two-dimensional, xi, 136, 189
Spectral radius, 55
Squid Loligo, 141
Stability, 44, 57, 79, 80
  analysis, 98
  bizarre, 252
  mean-square asymptotic, 74, 93, 250
Stability problem, 79
Stable,
  asymptotically, 39
  numerically, 39
State, 24
  time-asymptotic, 250
State equation, 22
State law, 23, 25, 27
Statics, 164
Stationary point, 53
Steady state, xiv, 250
  eventual, 93

Stellar interiors, 72
Stevens, Robert, xii
Stimulus, level of, 141
Stokes rule, 187, 188, 189
Stone–Weierstrass Theorem, 84
Stöss–Duhamel procedure, 189
Streamfunction, 26, 257
    equation, 26, 27, 77, 255
    potential, 250
Streamlines, 25, 26
Stress, 67, 170
String, taut, 254
Subdeterminants, 22
Sublinear case, 225
Subrectangle, 41
    closed, 114
    finite, 108
Subsonic flows, 27, 95, 255
Subspace, 223, 246
    linear, 236
Subthreshold stimuli, 135
Successively improved bounds, 109
Sufficient conditions, 35
Sufficiently smooth,
    data, 194
    function, 91, 184
    segment, 76
Summation convention, Einstein, 82
Sum,
    order of, 143
    partial, 142, 143
Superposition, 10
Supersonic flow, 27, 95
Supersonic nozzle, 260
Supersonic region, 77
Supremum, 227
Surface,
    equipotential, 255
    lateral, 67
Surface force, 66, 67
Systems,
    of first-order equations, 10
    of linear equations, 217

T

Tangents, to characteristics, 21
Tangents, vertical, 12
Taylor coefficients,
    positive, 152
    real, 49

Taylor series, 48, 146
    convergent, 57
Telegraphists' equation, 127
Telephony, xiii, 260
Telescoping, 138
Television antenna, 164
Temperature, 25
Term free of integration, 68
Terminating decimals, 38
"Test" functions, 91, 93
Tetrahedron, 88
Thermodynamics, 22, 25
Thickness parameter, 60
Thompson, Sir William, 258
Thompson principle, 174
Thrust, 169
Time, 131
Time-dependent flow, 22, 27
Time-dependent problems, 93, 250
Time-independent flow, 24, 27
    axially symmetric, 77
Timelike direction, positive, 80
Time to maximum effect, 135
Time trajectory, 66
Topology, algebraic, 88, 89
Toroidal circulation patterns, 141
Torque, per unit angle of twist, 170
Torsional rigidity, xiii, 170, 171, 174
Torsion, pure, 170
Total energy, 25
Traces of pressure waves, 23
Transformation, 243
    analytic, 147
    continuous linear, 243
    linear, 204
Transients, 93, 262
Transmission line equation, 22, 23, 131
Transmission lines, 64, 76, 133, 137, 259
    balanced, xiii
    theory, 72, 98, 131
Transoceanic communications, 131
Transonic flow, 95
Transonic speeds, controlled flight, 257
Trapezoids, 87
    finite union of, 87
Treatise on mechanics, 254
Triangular prisms, 88
Tricomi,
    equation, 16, 94, 95
    problem, 94
Trigonometric series, 39

Tube flow, 68
  large cylindrical, 253
  regular, 71
Tubes, with compliant walls, 93
Turbulence, phenomenon of, 260, 263
Type, independent of, 99, 146, 199
Types of equations,
  elliptic, 15, 16, 17, 27, 40, 42, 48, 57, 72, 78, 82, 83, 94, 154, 155, 249, 254, 255
  elliptic–parabolic, xi, xiii, 72, 82, 83, 93, 95, 199, 203, 205, 207, 210, 216, 225, 234, 240, 246, 247, 248
  hyperbolic, 9, 15, 16, 17, 19, 21, 22, 23, 27, 31, 40, 42, 48, 72, 76, 94, 97, 133, 136, 155, 166, 254
  mixed, elliptic–hyperbolic, 94, 95
  parabolic, 15, 16, 17, 22, 27, 40, 59, 65, 71, 72, 81, 94, 133, 254

U

Unique,
  continuation, 65, 255
  continuous extension, 14, 37
  limit, 64
  not, 13
  up to a constant, 52
Uniqueness, xi, xiii, xiv, 7, 8, 21, 36, 37, 40–43, 45, 51, 52, 57, 60, 63–65, 68, 69, 74, 84–87, 97, 98, 101, 102, 109–111, 117, 118, 121, 128, 149, 159, 160, 164, 165, 177, 178, 185, 200, 214, 226, 245, 248, 253–255
  mean square asymptotic, 69, 71, 176, 250, 262
  of regular solutions, 215, 216
  time asymptotic, xiv
  weakened sense of, 93, 262
Uniqueness problem, 8, 42, 60, 216
Universalist, 73
Unknowns, 55

V

Variables,
  complex, 16, 47, 49, 54, 254
  dependent, 95
  independent, 8, 11, 21, 22, 95, 97, 148
  of integration, 104
  real, 40, 222
Variational principles, xiii, 92, 169, 172

Vector field, 69
Vector space, 222, 225
  linear, 170, 221
Vectors,
  addition and subtraction, 222
  infinite-dimensional, 221
  multiplication by a scalar, 222
Vertebrate nerve fibres, 141
Vibrating elastic solid, 177
Vibrating string problem, 36, 37, 39, 40, 42, 59, 87, 257
  physical case, 39
Vibration, displacement, 254
Virtual mass, 171, 174
Viscosity coefficient, 27, 67
Viscous effects, 27, 255
Viscous flow, 27
  in a rigid pipe, 93, 176
Viscous shear, 66, 69, 253
Voice, 139
Voltage, 23, 24, 131, 132, 134, 136, 259
  drop, 131
Voltage induced, 24, 258
Voltage, ohmic, 24
Volume flow rate, 68, 169

W

Wall compliance, 67
Wave, 254
  advancing, 195
  character, 254
  front, 194
  motion, 259
  operator, 9, 137, 177
  propagation, 136, 137
  reflected, 196
Wave equation, xi, 16, 39, 41, 42, 43, 44, 45, 51, 56, 57, 94, 95, 97, 127, 136, 145, 177, 178, 185, 189, 193
  damped or undamped, 127
  in higher-dimensional space, 157
  homogeneous, 3, 29, 75, 78, 84, 101, 186, 193, 195, 197
  of Huyghens' type, 193
  nonhomogeneous, 68, 78, 94, 101, 118, 188, 192, 193, 194, 200
  nonhomogeneous (nonlinear), 114
  spherically symmetric, 197
Wave equation in E^2, 3, 29
Wavelengths, 24

Weakened sense of uniqueness, 74
Weierstrass, 145, 146
Weinberger, H. F., 176, 261
Weinstein, A., 174, 176
Well-posed problems, 18, 73, 75, 76, 77, 78, 81, 83, 84, 97

Weyl, Hermann, 261
Wire diameter and weight, 24, 40, 260
Woodhouse, Orange Wildman, 258

Z

Zorn, Max, 262